Lecture Notes in Mathematics

Edited by A. Dold and B. Eckmann

1377

John F. Pierce

Singularity Theory, Rod Theory, and Symmetry-Breaking Loads

Springer-Verlag

Berlin Heidelberg New York London Paris Tokyo

Author

John F. Pierce
Department of Mathematics
United States Naval Academy
Annapolis, MD 21402, USA

Mathematics Subject Classification (1980): 58F14, 73C50, 58C27, 58F05

ISBN 3-540-51304-3 Springer-Verlag Berlin Heidelberg New York
ISBN 0-387-51304-3 Springer-Verlag New York Berlin Heidelberg

Printing and binding: Druckhaus Beltz, Hemsbach/Bergstr.
2146/3140-543210

TABLE OF CONTENTS

I. INTRODUCTION

Take an initially straight rod of circular cross section which is composed of an isotropic material. Apply an axially symmetric compressive load to it. In general, the rod will assume an equilibrating configuration. However, this equilibrium is not isolated. Because of the symmetry of the rod and the load, the configuration will determine an entire family or "orbit" of other equilibrating configurations which is gained by rotating the image of the original configuration about the axis of symmetry through any angle.

Now perturb the compressive load by additional loads which break the axial symmetry. How does this perturbation alter the orbit of equilibrating configurations for the unperturbed problem? Will the perturbed problem also have an orbit of equilibrating configurations which is in bijective correspondence with the original orbit, or will the original orbit break or disappear with the perturbation? To what extent does the alteration of the original orbit depend upon the material comprising the rod? To what extent does it depend upon the manner in which the perturbing load breaks the axial symmetry of the original load?

This work indicates how we can address these questions using the methods of modern analysis and the theory of singularities or bifurcations in the presence of symmetry. We direct the work towards two groups of researchers: mathematicians and mechanicians. For the mathematician it illustrates how these tools can contribute greatly towards resolving problems of current interest in mechanics. Conversely, the rod problem gives the mechanician a concrete context in which to learn how to apply these nonlinear mathematical tools.

Generally speaking, we can isolate three aspects of the rod problem we've proposed. First, there is a "symmetric buckling" problem. It manifests itself in the buckling of the rod under axially-symmetric compression. Second, there is a "pure orbit-breaking" problem. It occurs prior to the buckling of the rod, when we fracture the orbit of equilibrating configurations to the symmetric problem by applying an asymmetric perturbing load. Finally, there is the "full" or "coupled" problem. We compress the rod sufficiently to buckle it, while simultaneously exerting perturbing loads which break the axial symmetry.

The analysis of the symmetric buckling problem has its origins in Euler's study of the elastica, with the establishment of the existence of a positive compression p_0 at which buckling of a centerline of a rod first occurs. By 1976 S. Antman and his associates had begun the analysis of the buckling of a nonlinear rod with circular cross section under axially-symmetric compression (see [5], [9]). They used the bifurcation theory of Crandall and Rabinowitz. In 1985, E. Buzano, G. Geymonat, and T. Poston applied to S. Antman's model the symmetric bifurcation theory developed by M. Golubitsky and D. Schaeffer [23] to study the buckling of a prismatic rod subject to axially symmetric compression [4]. In each case there is a distinguishing mathematical feature of the development. While the equations governing the behavior of the rod maintain their symmetry as the load increases, equilibrating configurations arise which exhibit less symmetry than the equations.

The "pure orbit-breaking" problem differs from the first problem in that the perturbing load actually breaks the symmetry of the equations governing the behavior of the rod. The perspective which we wish to use to analyze it was developed in 1983 by D. Chillingworth, J. Marsden, and Y. Wan in [1, 2, 3] to study Stoppelli's Problem, which is a problem in the three-dimensional theory of elasticity where analogous questions arise. Their contribution was to reformulate the symmetry-breaking problem as a problem of bifurcation defined on a manifold which was a group. The formulation then allows us to use the theory of singularities and bifurcation to carry out the analysis.

The results for these two aspects suggest that we may analyze the full problem by formulating it as a bifurcation problem on a space which is a semidirect product of a group and a vector space on which the group acts by means of a representation.

In this work we take the first step towards analyzing the full problem. First, we formulate as bifurcation problems all three aspects of the rod problem, under the assumptions of the Kirchhoff rod theory. We present the model for the rod in the Kirchhoff theory, formulate the equilibrium problem in a variational setting, and extract from the equilibrium problem the pure orbit-breaking problem, the pure symmetric buckling problem, and the full problem as bifurcation problems.

We then adapt the approach of [1, 2, 3] to analyze the pure orbit-breaking problem for the rod. The analysis predicts how an orbit of equilibrating configurations generated by the straight configuration for a rod in the Kirchhoff theory subject to an axially symmetric compressive load alters when we perturb the given load by dead loads which break the axial symmetry. The methods lead to a classification of the perturbing loads. For each type of perturbing load in the classification we determine whether or not the orbit of equilibrating configurations for the unperturbed problem breaks. If it does, we determine qualitatively how the orbit alters. We also determine whether or not the alteration depends upon the material comprising the rod. If it does, we examine whether the alteration is determined by the first-order (linear) approximation to the response of the material, or by the higher-order (nonlinear) approximations to the response. We illustrate these conclusions using specific perturbing loads.

We then comment on how we may use the notion of "unfoldings" in the theory of bifurcation to begin to analyze the full problem. The comments indicate what obstructions we encounter when we try to carry out the analysis. We close by summarizing the open questions, and by indicating some avenues for further investigation.

To apply the methods of the singularity theory we must formulate the rod problem in the language of modern analysis.

Consequently, in Sections 2 through 4 we develop in detail the equilibrium problem for a rod in the special Cosserat and the Kirchhoff theories as a problem involving a mapping between manifolds of functions. The formulation extends to a global setting the analytical formulation of the equilibrium problem for the two theories given by S. Antman in [9]. The models are examples of the general Hamiltonian structure for a rod model in the convective representation of [33] §6.

In Section 2 we present the geometry of the manifolds of configurations for a rod in the two theories. In Section 3 we specify the kind of perturbing loads we will examine (Assumption 3.2), and present the geometry of the spaces of admissible loads for each of the two theories. In Section 4 for each theory we show how the system of ordinary differential equations specifying an equilibrium configuration for the rod determines a mapping from the manifold of configurations into the space of loads (Theorem 4.20). Restricting attention to the Kirchhoff theory and assuming the material comprising the rod is hyperelastic, we extract from the equilibrium problem the general symmetry-breaking problem of interest as a problem of finding the singularities of a function (Problem 4.25), which is a particular type of bifurcation problem.

The development in Sections 2 and 3 is directed prinicipally towards mechanicians. Readers interested principally in the analysis of the mathematical model, and not in its development may pass directly to Section 4 with only a modicum of discomfort.

In Sections 5 through 7 we analyze Problem 4.25. In Section 5 we reduce the problem from one defined on a function space to one which is specified on a finite dimensional space. How the problem reduces depends upon the pressure of the compressive load. We examine the reduction for two cases: when the pressure approaches the value at which the rod first begins to buckle (Theorem 5.18), and when the pressure is at the first buckling value (Theorem 5.24). From the two reductions we formulate three finite-dimensional bifurcation problems (Problems 5.20, 5.26, and 5.27). They constitute the three aspects of the rod problem we presented at the beginning of the introduction.

We then restrict attention to the analysis of the pure orbit-breaking problem (Problem 5.20). We use the orbit generated by the straight configuration of the rod as the trivial orbit of equilibrating configurations for the bifurcation problem. How we proceed to resolve the reduced problem depends upon the nature of the perturbing load. In Section 6 we produce a classification for the perturbing loads (Theorem 6.20). In Section 7 we determine how the trivial orbit of equilibrating configurations for the symmetrically loaded problem alters for each type of perturbing load arising from the classification theorem (Theorems 7.6, 7.10, and 7.11). Some types of perturbing loads alter the orbit in a way which is independent of the material comprising the rod. Other types alter the orbit in ways which are determined by the first-order (linear) approximation of the response of the material comprising the rod, where the approximation is taken relative to the trivial configuration. Still other types of perturbing loads alter the orbit in ways which are determined by the higher-order (nonlinear) approximations to the response of the material comprising the rod taken relative to the trivial configuration.

To aid the comprehension we illustrate the principal results of Sections 3, 6 and 7 using specific perturbing loads. In Section 3 we present a variety of three-dimensional force distributions which produce the kind of rod loads we are admitting as perturbations (Examples 3.18-3.21). In Section 6 we illustrate the classification theorem by classifying the rod loads that were presented in Section 3 (Examples 6.21-6.24). In Section 7 we illustrate the various conclusions about how the orbit alters and what factors influence the alteration using the specific rod loads which were classified in Section 6 (Examples 7.14-7.17). Example 7.17 and the subsequent remarks are particularly worthy of note.

In Section 8 we comment on how we may begin to analyze the full problem (Problem 5.26) using the notion of unfoldings in the bifurcation theory. The comments indicate what obstructions we encounter using this perspective, and ways by which we may overcome them. We close by some other problems which may be investigated using the singularity theory.

The author gratefully acknowledges the illuminating comments of Drs. S. Antman and D. Chillingworth in correspondences and conversations throughout the development of this work. He also acknowledges the timely comments of Drs. J. Marsden, G. Geymonat, and T. Healey.

Finally, the author acknowledges the Faculty Senate of West Virginia University for providing financial support to help initiate this work. He also acknowledges the Departments of Mathematics at the University of Maryland and the United States Naval Academy for their hospitality and assistance in the completion of the manuscript.

II. THE SPACES OF CONFIGURATIONS

In this section we specify the spaces of configurations and strains for the rod in the special Cosserat and the Kirchhoff theories. We obtain kinematic models for the two rod theories in the spatial and convective representations of [33]. We relate their descriptions to the more classical descriptions in terms of director vectors. We examine some geometric features of the spaces which will be of importance in the latter sections. Finally, we generalize the differentiability class of the configurations.

II.1. *The Spaces of Classical Configurations for a Rod*

Fix an origin and a triad $\{ \underline{e}_j \mid j = 1, 2, 3 \}$ of orthonormal vectors in the physical space E^3. View a rod as a slender three-dimensional body $\mathcal{B}$ whose image in a reference configuration is a circularly cylindrical solid of length 1 and radius R, and whose line of centroids lies along the $\underline{e}_3$ axis with its left end at the origin. Identify a material point p in $\mathcal{B}$ with its coordinates $X = (X^1, X^2, S)$ in the reference configuration. Identify the parameter S, $0 \le S \le 1$, with the corresponding point $(0,0,S)$ on the line of centroids, or centerline for the rod. For S fixed, call the planar surface

$$\mathcal{B}(S) = \{ X = (X^1, X^2, S) \mid (X^1)^2 + (X^2)^2 \le R^2 \}$$

the *material cross section* for the rod at the point S on the centerline.

As in [6] view a rod theory as describing the behavior of a constrained three-dimensional material body. Assume that a configuration such a constrained body can attain is determined by specifying a position vector function x(S) and a pair of orthonormal vector functions $\underline{d}_\alpha(S)$, $\alpha = 1, 2$. Interpret x(S) as specifying the position of the centerline points in the new configuration, and interpret the $\underline{d}_\alpha(S)$ as determining the orientation of the plane of the section $\mathcal{B}(S)$ in the new configuration and a line in the plane.

The relation expressing how x and the $\underline{d}_\alpha$ specify a configuration $\mathfrak{X}$ for the three-dimensional body can be quite general (see [6], p. 323). For convenience, assume a particularly simple, but acceptable relation:

$$\mathfrak{X}(X^1,X^2,S) = x(S) + \phi^\alpha(X^1,X^2,S)\underline{d}_\alpha(S), \qquad (2.1)$$

where the summation is implied. The development we present remains valid for more general expressions.

As (2.1) indicates, the vector functions characterize those three-dimensional configurations for the rod for which the centerline may flex, twist, and elongate, and the cross sections for the rod may rotate and shear relative to the centerline. However, the cross sections remain planar and undistorted in shape.

From the three-dimensional invertibility condition ([6], p. 312) require

$$x'(S)\cdot\underline{d}_1(S) \times \underline{d}_2(S) \neq 0, \qquad (2.2)$$

or that the tangent to the centerline not lie in the plane of the cross section in any configuration.

For x, $\underline{d}_\alpha$ satisfying (2.2), define

$$\underline{d}_3(S) = \mathrm{sgn}(x'(S)\cdot\underline{d}_1(S) \times \underline{d}_2(S))\underline{d}_1(S) \times \underline{d}_2(S). \qquad (2.3)$$

Then $\{\underline{d}_j |\ j = 1, 2, 3\}$ form an orthonormal triad of vector functions, and

$$x'(S)\cdot\underline{d}_3(S) > 0. \qquad (2.4)$$

Call a collection $\{ x, \underline{d}_j, j = 1, 2, 3 \}$ of C^k vector functions satisfying (2.2) through (2.4) a C^k *configuration* for the rod in the *special Cosserat theory.*

A particular case of (2.2) is the requirement

$$| x'(S) \cdot \underline{d}_1(S) \times \underline{d}_2(S) | \equiv 1. \tag{2.5}$$

As (2.1) indicates, vector functions satisfying (2.5) describe configurations for the rod in which the centerline is inextensible, and cross sections do not shear relative to the centerline. Call (2.5) the *Kirchhoff hypothesis,* and call a collection $\{ x, \underline{d}_j, j = 1, 2, 3 \}$ of C^k vector functions satisfying (2.3) through (2.5) a C^k*configuration* for the rod in the *Kirchhoff theory.* Notice that $x'(S) = \underline{d}_3(S)$ for such a configuration.

We now represent the space of all C^k configurations for the rod in either theory in a manner which will allow us to take advantage of the elements of modern analysis. First, we simplify the description of a rod configuration by introducing orthogonal transformation-valued functions.

2.1 Lemma. *Let k be an integer, $k \geq 1$.*

a) Let $O(E^3)$ be the space of orthogonal linear transformations of E^3 for the given origin. Let $\gamma(S) \in O(E^3)$ be a C^k function on $I = [0,1]$. Then γ determines a unique C^k configuration in the Kirchhoff theory for which the left end of the centerline for the rod is fixed at the origin, and conversely.

b) Let $\chi = (x,\gamma)$, where $x(S) \in E^3$ and $\gamma(S) \in O(E^3)$ are C^k functions on I satisfying

$$\gamma(S)\underline{e}_3 \cdot x'(S) > 0. \tag{2.6}$$

Then χ determines a unique C^k configuration in the special Cosserat theory, and conversely.

Proof.

a) Given γ, define $x(S)$ and $\underline{d}_j(S)$ by requiring

$$\underline{d}_j(S) = \gamma(S)\underline{e}_j, \ j = 1, 2, 3, \tag{2.7}$$

$$x(S) = \int_0^S \underline{d}_3(t)dt = \int_0^S \gamma(t)\underline{e}_3 dt. \tag{2.8}$$

Then $x'(S) = \underline{d}_3(S)$, and conditions (2.3) through (2.5) follow. Also, $x(0) = 0 \in E^3$. Conversely, given (2.3) through (2.5) and $x(0) = 0$, we can solve (2.7) uniquely for $\gamma(S)$. Since $x'(S) = \underline{d}_3(S)$, (2.8) follows. If the vector functions are C^k, then γ is also.

b) Given $\chi = (x,\gamma)$, define the vector functions by (2.7). By (2.6), the vector functions will satisfy (2.2) through (2.4). Conversely, given the vector functions, (2.7) specifies χ. If the vector functions are C^k, then χ is also. ∎

2.2 Definition. *Let $k \geq 1$, and let $I = [0,1]$.*

a) Take the space of C^k configurations in the Kirchhoff theory for a rod with its left end fixed at $0 \in E^3$ to be

$$M = C^k(I, O(E^3)).$$

b) Take the space of C^k configurations for a rod in the special Cosserat theory to be

$$N = \{\ \chi = (x,\gamma) \in C^k(I, E^3 \times O(E^3)\ |\ (2.6) \text{ holds } \}. \quad ∎$$

2.3 Lemma. *For $k \geq 1$, M and N are differentiable manifolds.*

Proof.

a) By [7] § 4.4, $O(E^3)$ is a closed submanifold of the Banach space $L(E^3)$. By [8] § 13, it then follows that M is a closed submanifold of the Banach space $C^k(I, L(E^3))$.

b) From part a), $C^k(I, E^3 \times O(E^3)) = C^k(I,E^3) \times C^k(I,O(E^3))$ is a product manifold. Condition (2.6) characterizes N as an open set in this manifold; hence, it is a differentiable manifold. ∎

Remark. The Euler angles are a coordinate system on a portion of $O(E^3)$. By [8], p. 50, C^k curves of Euler angles can be used to construct a coordinate system over a portion of $C^k(I,O(E^3))$. It is this coordinate representation on M which is used in [4], [9], and [10] in their studies of rod problems.

We close the subsection by identifying groups of transformations which act effectively on M and N.

2.4 Definition. *Let*

$$SO(3) = \{ Q \in O(E^3) \mid \det Q = +1 \},$$
$$O(2) = \{ Q \in O(E^3) \mid Q\underline{e}_3 = \underline{e}_3 \},$$
$$SO(2) = O(2) \cap SO(3).$$

Let R_π, $J \in O(2)$, $\Sigma_3 \in O(E^3)$ *be defined by*

$$R_\pi \underline{e}_\alpha = -\underline{e}_\alpha,$$
$$J\underline{e}_1 = -\underline{e}_1, \quad J\underline{e}_2 = \underline{e}_2,$$
$$\Sigma_3 \underline{e}_\alpha = \underline{e}_\alpha, \quad \Sigma_3 \underline{e}_3 = -\underline{e}_3,$$

for $\alpha = 1, 2$. *Denote by* $\langle\Sigma_3\rangle$ *and* $\langle J\Sigma_3\rangle$ *the subgroups generated by the elements enclosed in the brackets (or equivalently, the smallest subgroups of* $O(E^3)$ *containing the elements enclosed in the brackets. Set*

$$G_s = \langle O(2), \Sigma_3\rangle,$$
$$\Gamma = G_s \cap SO(3),$$
$$\Pi_1 = O(E^3) \times G_s,$$
$$\Pi = O(E^3) \times \Gamma,$$
$$\Pi_s = SO(3) \times G_s,$$

where the latter three groups are external direct product groups, and the latter two groups are subgroups of Π_1.

a) *For* $g = (Q_1, Q_2) \in \Pi_1$, *for* $T \in L(E^3)$, *define* $g\cdot T \in L(E^3)$ *by*

$$g\cdot T = Q_1 T Q_2^T.$$

b) For $g \in \Pi_1$, $\gamma \in M$ and $\chi = (x,\gamma) \in N$, define $g\gamma \equiv \mathcal{T}_g\gamma \in M$ and $g\chi \equiv \mathcal{T}_g\chi \in N$ by

$$(\mathcal{T}_g\gamma)(S) = g\cdot\gamma(S) \tag{2.9}$$

$$(\mathcal{T}_g\chi)(S) = (Q_1 x(S), g\gamma(S)). \blacksquare \tag{2.10}$$

2.5 Lemma. *Π_1 acts on M and N as a group of transformations.*

a) Π_1 acts effectively on N, in that $\mathcal{T}_g\chi = \chi$, $\forall\chi \in N$ iff $g = (1,1) \in \Pi_1$.

b) Let

$$H = \{\ g \in \Pi_1 \ |\ \mathcal{T}_g\gamma = \gamma,\ \forall\gamma \in M\ \}.$$

Then $H = \langle(-1,-1)\rangle$.

c) Π and Π_s each act effectively on M.

Proof. Equations (2.9) and (2.10) imply that Π_1 acts as a group of transformations on M and N. Since $O(E^3)$ acts effectively on $\mathbb{R}^3$, the equations also imply that Π_1 acts effectively on N.

b) Let $g = (Q_1,Q_2)$. By choosing in turn $\gamma(S) \equiv Q_1^T$ and $\gamma(S) \equiv Q$ fixed, but arbitrary, (2.9) implies $g \in H$ if and only if $Q_1 = Q_2$ and Q_1 commutes with all elements of $O(E^3)$, or equivalently, belongs to the center of $O(E^3)$. As $\langle -1\rangle$ is the center of $O(E^3)$ (see [7], chapter 4), part b) follows.

c) As H is a normal subgroup of Π_1, the factor group Π_1/H acts effectively on M under the induced action. Hence, it suffices to show that Π_1 is isomorphic to the direct sums

$$\Pi_1 \approx \Pi \oplus H \approx \Pi_s \oplus H.$$

We establish the first direct sum. The second follows by an analogous argument. As H constitutes the center of Π_1, the subgroups Π and H commute. Since Γ is contained in SO(3), the intersection of Π and H is trivial. So $\Pi \oplus H$ is a subgroup of Π_1. Finally, for $g = (Q_1,Q_2) \in \Pi_1$, $\Phi g = \big((\det Q_1)g, (\det Q_1)(1,1)\big)$ specifies a group homomorphism of Π_1 onto $\Pi \oplus H$, giving the isomorphism. The induced action of the factor group then may be identified with the action of Π on M given by (2.9). $\blacksquare$

The action of Π on M given by Lemma 2.5 has a physical interpretation. Let $g = (Q_1, Q_2) \varepsilon \Pi$. the first factor corresponds to the usual action of $O(E^3)$ on the rod in space. If $Q_2 \in SO(2)$, the second factor rotates each cross section of the rod about the tangent to the centerline $\underline{d}_3(S) = x'(S)$ (see [9], p. 296). If $Q_2 = J\Sigma_3$, the second factor "flips" the cross sections by rotating each of them π radians about the $\underline{d}_2(S)$ axis.

II.2. *The Spaces of Infinitesimal Displacements*

In this subsection we identify the space of C^k infinitesimal displacements from a given configuration in each of the rod theories geometrically as the tangent space to the manifold of configurations at that configuration. We then identify the space of vector functions which produces the infinitesimal displacements with the tangent space to the manifold of configurations at the reference configuration. The identification produces the convective and spatial representations for the space of infinitesimal displacements of [33] for each of the rod theories. Finally, we describe the action of the group Π on the spaces.

In the special Cosserat theory a C^k infinitesimal displacement from a configuration $\{x, \underline{d}_j \mid j = 1, 2, 3\}$ is specified by C^k vector functions $\{ \delta x, \delta\underline{d}_j \mid j = 1, 2, 3 \}$ which satisfy the requirements

$$\delta\underline{d}_i(S)\cdot\underline{d}_j(S) + \underline{d}_i(S)\cdot\delta\underline{d}_j(S) \equiv 0, \tag{2.11}$$

for $i, j = 1, 2, 3$ ([11], [12]). It follows that there is a C^k vector function $z(S)$ for which

$$\delta\underline{d}_j(S) = z(S) \times \underline{d}_j(S). \tag{2.12}$$

In the Kirchhoff theory, if the $\delta\underline{d}_j$ satisfy condition (2.11) or (2.12), they specify a C^k infinitesimal displacement from a given configuration for the rod. The left end of the rod remains fixed at the origin, since (2.8) implies

$$\delta x(S) = \int_0^S \delta \underline{d}_3(t)\,dt.$$

The infinitesimal displacements may be viewed geometrically as elements of the tangent space at the appropriate configuration to the manifold of configurations of Definition 2.2. To construct the tangent space we use some information about the geometry of $O(E^3)$ and its consequences for M.

2.6 Lemma. *Let*

$$skew(E^3) = \{ W \in L(E^3) \mid W^T = -W \} .$$

For $a \in E^3$, *define* $\hat{a} \in skew(E^3)$ *by*

$$\hat{a}v = a \times v, \tag{2.13}$$

for $v \in E^3$.

a) The map $a \longrightarrow \hat{a}$ *is an isomorphism.*

b) For $A, B \in L(E^3)$, *define* $A{:}B = trace(AB^T)$. *Then for* $a, b \in E^3$,

$$a\cdot b = (1/2)\hat{a}{:}\hat{b}.$$

c) If $Q \in O(E^3)$, $a \in E^3$, *then*

$$\widehat{Qa} = (\det Q) Q\hat{a}Q^T \equiv (\det Q) Aut_Q \hat{a}.$$

d) If $a, b \in E^3$, *then*

$$\widehat{a \times b} = [\hat{a},\hat{b}] \equiv \hat{a}\hat{b} - \hat{b}\hat{a},$$

and
$$\widehat{a \times b} = b \otimes a - a \otimes b,$$

for
$$(a \otimes b)v \equiv (b\cdot v)a,$$

and $v \in E^3$.

Proof. See [1], p. 300. ■

2.7 Lemma. *$O(E^3)$ is a Lie group with left and right translations given by*

$$\mathcal{L}_Q Q_1 = \mathcal{R}_{Q_1} Q = QQ_1 .$$

a) For $Q \in O(E^3)$, the tangent space to $O(E^3)$ at Q is

$$T_Q O(E^3) = \{ H \in L(E^3) \mid HQ^T + QH^T = 0 \}.$$

In particular, for $Q = 1$, the identity transformation,

$$T_1 O(E^3) = skew(E^3).$$

b) If $X, Y \in skew(E^3)$, set

$$[X,Y] = XY - YX .$$

Then $skew(E^3)$ is an algebra, the Lie algebra of $O(E^3)$.
c) Each $Q \in O(E^3)$ determines two isomorphisms of $skew(E^3)$ onto $T_Q O(E^3)$: for $W \in skew(E^3)$,

$$L_Q W \equiv T\mathcal{L}_Q(1)W \equiv QW,$$
$$R_Q W \equiv T\mathcal{R}_Q(1)W \equiv WQ.$$

Moreover, the isomorphisms vary smoothly with Q.

Proof. See [13], chapter 1 and [7], chapter 4. ■

In the manner of [28] M inherits a Lie group structure from $O(E^3)$.

2.8 Proposition. *For $\gamma, \eta \in M$,*
a) define $\gamma\eta \in M$ by

$$\gamma\eta(S) = \gamma(S)\eta(S) ,$$

where the right hand side denotes the product in $O(E^3)$. Then M with this product is a Lie group.

b) Define $\mathcal{L}_\gamma\eta \in M$ and $\mathcal{R}_\gamma\eta \in M$ by

$$(\mathcal{L}_\gamma\eta)(S) \equiv \mathcal{L}_{\gamma(S)}\eta(S) = \gamma(S)\eta(S)$$
$$(\mathcal{R}_\gamma\eta)(S) \equiv \mathcal{R}_{\gamma(S)}\eta(S) = \gamma(S)\eta(S) ,$$

where the right hand sides are the translations in $O(E^3)$. Then $\mathcal{L}_\gamma$ and $\mathcal{R}_\gamma$ are diffeomorphisms on M and constitute the left and right translations on M for the Lie group structure.

Proof. Since the multiplication is defined pointwise on I, and since $O(E^3)$ is a Lie group, we need only establish that the multiplication is smooth on M. But by [8] §13, when defined, multiplication of two C^k mappings is itself a smooth map. ∎

We can now geometrically characterize a C^k infinitesimal displacement.

2.9 Proposition.

a) Let $\gamma \in M$, and let $\delta\underline{d}_j$, $j = 1, 2, 3$ be a C^k infinitesimal displacement from the configuration determined by γ. For $S \in I$, define $H_\gamma(S) \in L(E^3)$ by

$$\delta\underline{d}_j(S) = H_\gamma(S)\underline{e}_j, \tag{2.14}$$

$j = 1, 2, 3$. Then

(1) $H_\gamma(S) \in T_{\gamma(S)}O(E^3)$,

(2) H_γ is C^k in S,

(3) $\delta\underline{d}_j(S) = z(S) \times \underline{d}_j(S)$, if and only if

$$H_\gamma(S) = \hat{z}(S)\gamma(S). \tag{2.15}$$

Conversely, if $\gamma \in M$, and if H_γ is a C^k function satisfying conditions (1) and (2), then (2.14) specifies a C^k infinitesimal displacement from γ in the Kirchhoff theory which satisfies condition (3).

b) Let $\chi \in N$, and let δx, $\delta \underline{d}_j$, $j = 1, 2, 3$ specify a C^k infinitesimal displacement from the configuration described by χ. For $S \in I$, define $h(S) \in E^3$ and $H_\gamma(S) \in L(E^3)$ by (2.14) and

$$\delta x(S) = h(S). \tag{2.16}$$

Then conditions (1) through (3) of part a) hold. Conversely, if $\chi \in N$, and if h and H_γ are C^k functions which satisfy conditions (1) and (2), then (h, H_γ) specify a C^k infinitesimal displacement from χ in the special Cosserat theory which satisfies condition (3).

Proof.

a) By (2.7), (2.11), and (2.14), $H_\gamma(S)$ satisfies Lemma 2.7 a) for $Q = \gamma(S)$, giving condition (1). Condition (2) follows from (2.14) and the differentiability assumption. Condition (3) follows from (2.13) and (2.14). Conversely, given H_γ, define $\delta \underline{d}_j$ by (2.14) and δx by

$$\delta x(S) = \int_0^S H_\gamma(t)dt \, \underline{e}_3.$$

Equations (2.11) and (2.14) then follow, giving the conclusion.

b) Conditions (1) through (3) follow as in part a). Conversely, given (h, H_γ), define $\delta x = h$ and $\delta \underline{d}_j$ by (2.14). Then condition (1) implies that (2.11) is satisfied, giving the conclusion. ∎

We can now relate the infinitesimal displacements to elements of the tangent spaces of M and N.

2.10 Proposition. *Let $\gamma \in M$ and $\chi = (x, \gamma) \in N$.*

a) Let $T_\gamma M$ and $T_\chi N$ denote the tangent spaces to M and N at γ and χ, respectively. Then

$$T_\gamma M = \{ H_\gamma \mid H_\gamma(S) \in T_{\gamma(S)}O(E^3),\ H_\gamma \text{ is } C^k \text{ in } S \},$$

$$T_\chi N = C^k(I, E^3) \times T_\gamma M.$$

In particular, if $\gamma_0 \in M$ *is defined by* $\gamma_0(S) \equiv 1 \in O(E^3)$, *and if* $\chi_0 \in N$ *is defined by* $\chi_0 = (x_0, \gamma_0)$, $x_0(S) = S\underline{e}_3$, *then*

$$T_{\gamma_0} M = C^k(I, skew(E^3)),$$
$$T_{\chi_0} N = C^k(I, E^3 \times skew(E^3)).$$

b) *For* $V, W \in T_{\gamma_0} M$ *define* $[V,W] \in T_{\gamma_0} M$ *by*

$$[V,W](S) = [V(S), W(S)] ,$$

where the right hand side is specified in Lemma 2.7. Then $T_{\gamma_0} M$ *with this product is an algebra, the Lie algebra for the Lie group* M.

c) *Given* $\gamma \in M$ *there are two isomorphisms of* $T_{\gamma_0} M$ *onto* $T_\gamma M$ *which are left and right invariant, respectively. If* $Z \in T_{\gamma_0} M$, *they are:*

$$(1) \quad L_\gamma Z \equiv T\mathcal{L}_\gamma(\gamma_0)Z \equiv H_\gamma , \tag{2.17}$$

where $H_\gamma(S) = \gamma(S)Z(S)$,

$$(2) \quad R_\gamma Z \equiv T\mathcal{R}_\gamma(\gamma_0)Z \equiv H_\gamma , \tag{2.18}$$

where $H_\gamma(S) = Z(S)\gamma(S)$,

and the right hand sides are as specified in Lemma 2.7. The isomorphisms vary smoothly with γ.

d) *If* $\chi \in N$, *then there are two isomorphisms of* $T_{\chi_0} N$ *onto* $T_\chi N$ *which are left and right invariant, respectively. If* $(b,Z) \in T_{\chi_0} N$, *they are:*

$$(1) \quad L_\chi(b,Z) \equiv T\mathcal{L}_\chi(\chi_0)(b,Z) = (h, H_\gamma),$$

where $h(S) = \gamma(S)b(S)$, *and* H_γ *is given by (2.17),*

$$(2) \quad R_\chi(b,Z) \equiv T\mathcal{R}_\chi(\chi_0)(h,Z) = (h, H_\gamma),$$

where $h(S) = b(S)$, and H_γ is given by (2.18).

The isomorphisms vary smoothly with χ.

Proof. The assertions for part a) follow from [8], p. 51 and Lemma 2.7. The assertions for parts b) and c) follow from Lemma 2.7 and the α-Lemma ([8], p. 32) in the manner of [28]. ∎

2.11 Theorem.

a) If $\gamma \in M$, then $T_\gamma M$ is the space of all C^k infinitesimal displacements from the configuration γ in the Kirchhoff theory for the rod with its left end fixed.

b) If $\chi \in N$, then $T_\chi N$ is the space of all C^k infinitesimal displacements from the configuration χ in the special Cosserat theory for the rod.

Proof. The theorem follows from Propositions 2.8 and 2.9. ∎

The isomorphisms of Propositions 2.10, the way we characterize rigid body motion in terms of Lie groups ([7] §4.4,), and the models of [33] motivate the following terminology.

2.12 Definition. *Let*

$$\Delta(S) = \int_0^S \|x'(T)\| \, dT$$

denote the arclength function for the centerline of the rod in the given configuration in either of the two rod theories.

a) Let $C = T_{\gamma_0} M$ and $D = T_{\chi_0} N$. Call C and D the representation spaces *for the C^k infinitesimal displacements in the Kirchhoff and special Cosserat theory, respectively.*

b) Let $\gamma \in M$ and $H_\gamma \in T_\gamma M$.

(1) Call $Z \in C$ for which $L_\gamma Z = H_\gamma$ the convective representation *for H_γ.*

(2) Call $Z_2 \in C$ for which $\Delta'(S)(R_\gamma Z_2)(S) = H_\gamma(S)$ the spatial representation *for H_γ.*

c) Let $\chi \in N$ and $(h, H_\gamma) \in T_\chi N$.

(1) Call $(b,Z) \in D$ for which $L_\chi(b,Z) = (h,H_\gamma)$ the convective representation *for (h,H_γ).*

(2) Call $(b_2,Z_2) \in D$ for which $\Delta'(S)\big(R_\chi(b_2,Z_2)\big)(S) = \big((h,H_\gamma)\big)(S)$ the spatial representation *for (h,H_γ).* ∎

By Lemma 2.7 and Proposition 2.10, the two representations are related by the equations

$$\left.\begin{aligned} \Delta'(S)b_2(S) &= \gamma(S)b(S), \\ \Delta'(S)Z_2(S) &= \gamma(S)Z(S)\gamma^T(S) \equiv \mathrm{Aut}_{\gamma(S)}Z(S). \end{aligned}\right\} \quad (2.19)$$

The spatial representation presented in Definition 2.12 is identical to the classical description of a infinitesimal displacement given by (2.12). Specifically, if $\delta \underline{d}_j = z \times \underline{d}_j$, then $(b_2,Z_2) = (\delta x,\hat{z})$ is the spatial representation for the infinitesimal displacement. By (2.19), the convective representation for the infinitesimal displacement is specified in terms of its classical description by

$$(b,Z) = (\gamma^T\delta x,\gamma^T z). \quad (2.20)$$

Finally, in the usual presentation of the rod theory it is advantageous to represent the classical description of a infinitesimal displacement in terms of its components relative to the vector functions $\underline{d}_j$ ([11], [12]). If $\delta \underline{d}_j = z \times \underline{d}_j$, (2.7) and the above remarks imply

$$z(S)\cdot\underline{d}_j(S) = \gamma^T(S)z(S)\cdot\underline{e}_j,$$

and

$$\delta x(S)\cdot\underline{d}_j(S) = \gamma^T(S)\delta x(S)\cdot\underline{e}_j.$$

So (2.20) indicates that representing a infinitesimal displacement in this manner is equivalent to specifying its convective representation.

We conclude the subsection by indicating how the group Π_1, Π, and Π_s of Lemma 2.5 act on the infinitesimal displacements and their spatial and convective representations.

2.13 Proposition. *Let $g = (Q_1,Q_2) \in \Pi_1$, let $\gamma \in M$, and let $\chi \in N$. Let $\mathcal{T}_g$ denote the action of Π on M or N given by Definition 2.4.*

a) *If $H_\gamma \in T_\gamma M$, then*

$$T\mathcal{T}_g(\gamma)H_\gamma \equiv g\cdot H_\gamma \in T_{g\gamma}M,$$

and is given by

$$(g\cdot H_\gamma)(S) = Q_1 H_\gamma(S) Q_2^{\mathrm{T}} . \tag{2.21}$$

b) *If $H_\chi \equiv (h,H_\gamma) \in T_\chi N$, then*

$$T\mathcal{T}_g(\chi)H_\chi \equiv g\cdot H_\chi \in T_{g\chi}N ,$$

and is given by

$$(g\cdot H_\chi)(S) = (Q_1 h(S), Q_1 H_\gamma(S) Q_2^{\mathrm{T}}). \tag{2.22}$$

Proof.

a) As in [13], the action of Π on $O(E^3)$ induces an action on tangent spaces: for $H \in T_Q O(E^3)$,

$$g\cdot H = Q_1 H Q_2^{\mathrm{T}} \in T_{g\cdot Q} O(E^3).$$

For $\gamma \in M$, [8] implies that this action extends to $T_\gamma M$.

b) If $v \in E^3$, $g\cdot v = Q_1 v$ specifies an action of Π on E^3. By [8], this action extends to $C^k(I,E^3)$. Since $T_x C^k(I,E^3) = C^k(I,E^3)$ for $x \in C^k(I,E^3)$, the conclusion follows in the manner of part a). ∎

We can now determine how the convective and spatial representations for an infinitesimal displacement change as a result of the action of Π_1.

2.14 Proposition. *Let $g = (Q_1,Q_2) \in \Pi_1$, let $\gamma \in M$, and let $\chi \in N$.*

a) *Let $H_\gamma \in T_\gamma M$.*

(1) *If $Z \in C$ satisfies $L_\gamma Z = T\mathcal{L}_\gamma(\gamma_0)Z = H_\gamma$, then*

$$T\mathcal{T}_g(\gamma)H_\gamma \equiv g\cdot H_\gamma = T\mathcal{L}_{g\gamma}(\gamma_0)Ad_{Q_2}Z \ ,$$

where $Ad_{Q_2}Z \in C$ is given by

$$(Ad_{Q_2}Z)(S) = Ad_{Q_2}(Z(S)) \equiv Q_2 Z(S)Q_2^T \ .$$

(2) *If $W \in C$ satisfies $T\mathcal{R}_\gamma(\gamma_0)W = H_\gamma$, then*

$$T\mathcal{T}_g(\gamma)H_\gamma \equiv g\cdot H_\gamma = T\mathcal{R}_{g\gamma}(\gamma_0)Ad_{Q_1}Z \ ,$$

where $Ad_{Q_1}Z \in C$ is given by

$$(Ad_{Q_1}Z)(S) = Ad_{Q_1}(Z(S)) \equiv Q_1 Z(S)Q_1^T \ .$$

b) *Let $(h,H_\gamma) \in T_\chi N$.*
(1) *If $(b,Z) \in D$ satisfies $T\mathcal{L}_\chi(\chi_0)(b,Z) = H_\chi$, then*

$$T\mathcal{T}_g(\chi)H_\chi = T\mathcal{L}_{g\chi}(\chi_0)(Q_2 b, Ad_{Q_2}Z) \ .$$

(2) *If $(1,W) \in D$ satisfies $T\mathcal{R}_\chi(\chi_0)(1,W) = H_\chi$, then*

$$T\mathcal{T}_g(\chi)H_\chi = T\mathcal{R}_{g\chi}(\chi_0)(Q_1 1, Ad_{Q_1}W) \ .$$

Proof. Since

$$(Q_1\gamma(S)Q_2^T)^T Q_1 H_\gamma(S)Q_2^T = Q_2\gamma(S)^T H_\gamma(S)Q_2^T,$$

part a) (1) follows from Proposition 2.13 and Definition 2.12. The other parts follow similarly. ∎

Proposition 2.14 leads us to define the following actions of Π_1 on the representation spaces for the infinitesimal displacements.

2.15 Definition. *Let $g = (Q_1, Q_2) \in \Pi_1$.*

a) If $Z \in C$, define $\mathcal{B}(g)Z$, $\mathcal{B}_2(g)Z \in C$ by

$$\mathcal{B}(g)Z = Ad_{Q_2} Z \tag{2.23}$$

$$\mathcal{B}_2(g)Z = Ad_{Q_1} Z \tag{2.24}$$

b) If $(b,Z) \in D$, define $\mathcal{B}(g)(b,Z)$, $\mathcal{B}_2(g)(b,Z) \in D$ by

$$\mathcal{B}(g)(b,Z) = (Q_2 b, Ad_{Q_2} Z) \ ,$$

$$\mathcal{B}_2(g)(b,Z) = (Q_1 b, Ad_{Q_1} Z) \ .$$

When restricted to Π (respectively, Π_s), call $\mathcal{B}$ (respectively, $\mathcal{B}_2$) the convective *(*spatial*) representation of Π (Π_s) on the representation space C for the Kirchhoff model. Similarly, for the special Cosserat theory, call $\mathcal{B}$ and $\mathcal{B}_2$, respectively, the* convective *and* spatial *representation for Π_1 on the representation space D.* ∎

Remark. The restriction to Π or Π_s in Definition 2.15 for the Kirchhoff theory arises from the fact that Π_1 does not act effectively on M. See Lemma 2.5.

The various actions and representations are related.

2.16 Lemma. *Let $g \in \Pi_1$, $\gamma \in M$, and $\chi \in N$.*

a) If $Z \in C$, then

$$T\mathcal{R}_{g\gamma}(\gamma_0)(\mathcal{B}_2(g)Z) = T\mathcal{I}_g(\gamma)T\mathcal{R}_\gamma(\gamma_0)Z, \tag{2.25}$$

$$T\mathcal{L}_{g\gamma}(\gamma_0)(\mathcal{B}(g)Z) = T\mathcal{I}_g(\gamma)T\mathcal{L}_\gamma(\gamma_0)Z \ . \tag{2.26}$$

b) If $(b,Z) \in D$, then

$$T\mathcal{R}_{g\chi}(\chi_0)(\mathcal{B}_2(g)(b,Z)) = T\mathcal{I}_g(\chi)T\mathcal{R}_\chi(\chi_0)(b,Z) \ , \tag{2.27}$$

$$T\mathcal{L}_{g\chi}(\chi_0)(\mathcal{B}(g)(b,Z)) = T\mathcal{I}_g(\chi)T\mathcal{L}_\chi(\chi_0)(b,Z) \ . \tag{2.28}$$

Proof. The lemma follows from Definitions 2.12 and 2.15 and Proposition 2.14. ∎

II.3. *The Manifolds of Generalized and Constrained Configurations*

In section V we will use the theory of elliptic differential equations to examine an equilibrium problem for a rod in the Kirchhoff theory. Consequently, we introduce configurations for the rod whose differentiability class is more general than those in M or N.

2.17 Definition. *Let $k \geq 1$. Let r be an integer satisfying $r > k + (1/2)$. Set*

a) $\mathcal{M} = W^{r,2}(I, O(E^3))$,

b) $\mathcal{N} = \{\chi = (x,\gamma) \in W^{r,2}(I,E^3) \times \mathcal{M} \mid x'(S)\cdot\gamma(S) \neq 0,\ S \in I\}$.

Call $\mathcal{M}$ and $\mathcal{N}$ the manifolds of $W^{r,2}$ configurations *for the Kirchhoff and special Cosserat theories, respectively.* ∎

Here, x and γ in $W^{r,2}$ means x and γ and their derivatives up through order r have square integrable norms over I: for j = 1, 2, ... r,

$$\int_I x^{(j)}(S)\cdot x^{(j)}(S)\,dS < \infty,$$

$$\int_I \gamma^{(j)}(S):\gamma^{(j)}(S)\,dS < \infty.$$

The derivatives $\gamma^{(j)}$ are defined using the fact that $O(E^3)$ is a Riemannian manifold ([7] § 4.4).

Since $O(E^3)$ is a smooth, closed submanifold of $L(E^3)$, $\mathcal{M}$ is a smooth, closed submanifold of Banach space $W^{r,2}(I,L(E^3))$ (see [8] § 13). Since $r > k + (1/2)$, the Sobolev inequalities ([8], p. 26) imply that $\mathcal{M} \subseteq C^k(I,O(E^3))$ and $\mathcal{N} \subseteq C^k(I,E^3 \times O(E^3))$.

Consequently, the condition specifying $\mathcal{N}$ is well defined, and it follows that $\mathcal{N}$ is a smooth differentiable manifold.

We can identify the $W^{r,2}$ infinitesimal displacements with the elements of the tangent spaces to $\mathcal{M}$ and $\mathcal{N}$ in the manner of Proposition 2.10 and Theorem 2.11. However, to identify the convective and spatial representations of a $W^{r,2}$ infinitesimal displacement, we need a technical lemma.

2.18 Lemma. *Let $r \geq 1/2$.*

a) The multiplications

$$\left.\begin{array}{ccc} W^{r,2}(I,O(E^3)) \times W^{r,2}(I,O(E^3)) & \longrightarrow & W^{r,2}(I,O(E^3)) \\ \gamma,\ \eta & \longmapsto & \gamma\eta \end{array}\right\} \quad (2.29)$$

$$\left.\begin{array}{ccc} W^{r,2}(I,O(E^3)) \times W^{r,2}(I,E^3) & \longrightarrow & W^{r,2}(I,E^3) \\ \gamma,\ x & \longmapsto & \gamma x \end{array}\right\} \quad (2.30)$$

are smooth mappings.

b) The inversion map $\gamma \longmapsto \gamma^{-1} = \gamma^T$ is a smooth mapping on $W^{r,2}(I,O(E^3))$.

Proof.

a) See [8], p. 33.

b) In this case, the inversion map is the restriction of a linear map on the Banach space $W^{r,2}(I,L(E^3))$ to the closed submanifold $\mathcal{M}$. ∎

2.19 Corollary. *$\mathcal{M}$ with the multiplication of (2.29) is a Lie group. For γ, $\eta \in \mathcal{M}$ the mappings $\mathcal{L}_\gamma\eta \in \mathcal{M}$ and $\mathcal{R}_\eta\gamma \in \mathcal{M}$ given by*

$$\mathcal{L}_\gamma\eta = \mathcal{R}_\eta\gamma = \gamma\eta$$

constitute the left and right translations for the Lie group.

Proof. The differentiability conditions for the Lie group multiplication follow from Lemma 2.18. ∎

2.20 Proposition. *Let $k \geq 1$ and $r > k + 1/2$.*

a) If $\gamma \in \mathcal{M}$ and if $\chi = (x,\gamma) \in \mathcal{N}$, then

$$T_\gamma \mathcal{M} = \{H_\gamma \mid H_\gamma(S) \in T_{\gamma(S)}O(E^3),\ H_\gamma \text{ is a } W^{r,2} \text{ function}\}, \tag{2.31}$$

$$T_\chi \mathcal{N} = \{H_\chi = (h,H_\gamma) \mid h \in W^{r,2}(I,E^3),\ H_\gamma \in T_\gamma \mathcal{M}\}. \tag{2.32}$$

In particular, if $\gamma_0 \in \mathcal{M}$ is given by $\gamma_0(S) \equiv I \in O(E^3)$, and if $\chi_0 = (x_0,\gamma_0)$, $x_0(S) = S\underline{e}_3$,

$$T_{\gamma_0}\mathcal{M} = W^{r,2}(I, skew(E^3)),$$

$$T_{\chi_0}\mathcal{N} = W^{r,2}(I, E^3 \times skew(E^3)).$$

b) For $V, W \in T_{\gamma_0}\mathcal{M}$, define $[V,W]$ by

$$[V,W](S) = V(S)W(S) - W(S)V(S)\ .$$

then $[V,W] \in T_{\gamma_0}\mathcal{M}$, and $T_{\gamma_0}\mathcal{M}$ with this product is the Lie algebra for $\mathcal{M}$.

c) If $\gamma \in \mathcal{M}$, the map

$$\begin{aligned} L_\gamma:\ T_{\gamma_0}\mathcal{M} &\longrightarrow T_\gamma\mathcal{M} \\ Z &\longmapsto L_\gamma Z \equiv T\mathcal{L}_\gamma(\gamma_0)Z \end{aligned}$$

is given by $\qquad (L_\gamma Z)(S) = \gamma(S)Z(S)$

and is an isomorphism which varies smoothly with γ.

d) If $\chi = (x,\gamma) \in \mathcal{N}$, the map

$$\begin{aligned} L_\chi:\ T_{\chi_0}\mathcal{N} &\longrightarrow T_\chi\mathcal{N} \\ (b,Z) &\longmapsto L_\chi(b,Z) = (h,H_\gamma) \end{aligned}$$

is given by $\qquad h(S) = \gamma(S)b(S)$ *and* $H_\gamma = T\mathcal{L}_\gamma(\gamma_0)Z$

and is an isomorphism which varies smoothly with χ.

Proof.

a) $T_\gamma\mathcal{M}$ follows from [8], p. 50. $T_\chi\mathcal{N}$ follows from the fact that $\mathcal{N}$ is an open submanifold of $W^{r,2}(I,E^3)$ X $\mathcal{M}$.

b) Lemma 2.18 implies that $[V,W] \in T_{\gamma_0}\mathcal{M}$. The Lie algebra structure follows in the manner of the proof of Proposition 2.10 (see [28]).

c) By Corollary 2.19 $\mathcal{L}_\gamma$ is the left translation for the Lie group $\mathcal{M}$. Hence $T\mathcal{L}_\gamma(\gamma_0)$ is an isomorphism and varies smoothly with γ_0.

d) This part follows from part c) in the manner of Proposition 2.10 d). ■

2.21 Definition. *Let $\gamma \in \mathcal{M}$ and $\chi \in \mathcal{N}$. Set $\mathcal{C} = W^{r,2}(I,skew(E^3))$ and $\mathcal{D} = W^{r,2}(I,E^3 \text{ X } skew(E^3))$.*

a) If $H_\gamma \in T_\gamma\mathcal{M}$ call $Z \in \mathcal{C}$ for which

$$L_\gamma Z = H_\gamma$$

the convective representation for H_γ. Call $\mathcal{C}$ the representation space for the $W^{r,2}$ infinitesimal displacements in the Kirchhoff theory.

b) If $H_\chi = (h,H_\gamma) \in T_\chi\mathcal{N}$ call $(b,Z) \in \mathcal{D}$ for which

$$L_\chi(b,Z) = H_\chi$$

the convective representation for H_χ. Call $\mathcal{D}$ the representation space for the $W^{r,2}$ infinitesimal displacements in the special Cosserat theory. ■

We can extend the action of the group Π to $\mathcal{M}$ and $\mathcal{N}$. The extension induces an action of Π on the spaces of $W^{r,2}$ infinitesimal displacements and the representation spaces.

2.22 Lemma.

a) The actions of Π_1 on M and N defined by (2.9) and (2.10) extend to $\mathcal{M}$ and $\mathcal{N}$.

b) The induced actions of Π_1 defined by (2.21) and (2.22) extend to $T_\gamma\mathcal{M}$ and $T_\chi\mathcal{N}$.

c) The representations of Π, Π, or Π_s on C and D defined by (2.23) through (2.26) extend to $\mathcal{C}$ and $\mathcal{D}$. Moreover, the *representations are orthogonal with respect to the $W^{r,2}$ inner products on $\mathcal{C}$ and $\mathcal{D}$.*

d) Lemma 2.16 extends to the spaces of generalized differentiability.

Proof. If $g = (Q_1,Q_2)$, Lemma 2.18 implies that the map $\gamma \longmapsto Q_1\gamma Q_2^T$ is a smooth diffeomorphism of $\mathcal{M}$. Part a) then follows in the manner of Lemma 2.6. The other three parts follow from Lemma 2.17 in the manner of Lemma 2.16 and Propositions 2.13 and 2.14. Finally, the orthogonality of the action of Π on $\mathcal{C}$ and $\mathcal{D}$ follows from the definition of the $W^{r,2}$ inner product, and the fact that Π acts orthogonally on E^3 and $L(E^3)$. ∎

In the problem which we introduce in Section IV we will consider a rod in the Kirchhoff theory which is constrained in such a way that the right end of the centerline lies on the $\underline{e}_3$ axis,

$$x(1)\cdot\underline{e}_\alpha = 0, \ \alpha = 1, \ 2. \tag{2.33}$$

We incorporate the constraint into the specification of the manifold of configurations using (2.8). We specify the manifold both in terms of configurations which are C^k, and those which are $W^{r,2}$.

2.23 Definition. *Let $k \geq 1$ and let $r > k + 1/2$. Define the spaces of C^k and $W^{r,2}$ configurations for the Kirchhoff theory for the rod with left end fixed at the origin and right end on the $\underline{e}_3$ axis to be*

$$M_1 = \{ \gamma \in M \mid \int_I \gamma(S)dS:(\underline{e}_\alpha \otimes \underline{e}_3) = 0, \ \alpha = 1, \ 2 \},$$

$$\mathcal{M}_1 = \{ \gamma \in \mathcal{M} \mid \int_I \gamma(S)dS:(\underline{e}_\alpha \otimes \underline{e}_3) = 0, \ \alpha = 1, \ 2 \},$$

respectively. ∎

2.24 Proposition. *M_1 and $\mathcal{M}_1$ are smooth, closed submanifolds of M and $\mathcal{M}$, respectively. If $\gamma \in M_1$ ($\mathcal{M}_1$), the spaces of C^k and $W^{r,2}$*

infinitesimal displacements from γ compatible with the constraints are

$$T_\gamma M_1 = \{ H_\gamma \in T_\gamma M \mid \int_I H_\gamma((S)dS:(\underline{e}_\alpha \otimes \underline{e}_3) = 0, \ \alpha = 1, 2 \}, \tag{2.34}$$

$$T_\gamma \mathcal{M}_1 = \{ H_\gamma \in T_\gamma \mathcal{M} \mid \int_I H_\gamma((S)dS:(\underline{e}_\alpha \otimes \underline{e}_3) = 0, \ \alpha = 1, 2 \}, \tag{2.35}$$

respectively.

Proof. The map from M into R^2 given by

$$i(\gamma) = \left(\int_I \gamma(S)dS:(\underline{e}_\alpha \otimes \underline{e}_3) \right) \underline{e}_\alpha \tag{2.36}$$

is smooth, $i^{-1}(\underline{0}) = M_1$ by construction. Since i is the restriction to M of a linear map on $C^k(I,L(E^3))$, if $\gamma \in M_1$ and if $H_\gamma \in T_\gamma C^k(I,L(E^3))$,

$$[Ti(\gamma)]H_\gamma = \left(\int_I H_\gamma(S)dS:(\underline{e}_\alpha \otimes \underline{e}_3) \right) \underline{e}_\alpha \, . \tag{2.37}$$

If $\underline{0} \in \mathbb{R}^2$ is a regular value for i, that is , $Ti(\gamma)$ has maximal rank at each $\gamma \in i^{-1}(\underline{0}) = M_1$, then [14] p. 862 implies that M_1 is a closed submanifold of codimension 2. So assume there is a $\gamma \in M_1$ for which $Ti(\gamma)$ has rank ≤ 1. Without loss of generality, assume that $i_1(\gamma) := i(\gamma)\cdot\underline{e}_1$ has a derivative with rank zero at γ. Let $H_\gamma = T\mathcal{L}_\gamma(\gamma_0)Z$ for $Z \in C$. Then (2.37) and the assumption implies

$$\left.\begin{aligned} Ti_1(\gamma)T\mathcal{L}_\gamma(\gamma_0)Z &= \left[\int_I \gamma(S)Z(S):\underline{e}_1 \otimes \underline{e}_3 \right] dS = \\ &= \left[\int_I \gamma^T(S)\underline{e}_1 \otimes \underline{e}_3 : Z(S) \right] dS = \\ &= \left[\int_I \mathrm{skew}(\gamma^T(S)\underline{e}_1 \otimes \underline{e}_3):Z(S) \right] dS = 0. \end{aligned}\right\} \tag{2.38}$$

In particular, take $Z(S) = \mathrm{skew}(\gamma^T(S)\underline{e}_1 \otimes \underline{e}_3) \in C$. Then (2.38) implies

$$\gamma_{11}(S) = \gamma_{12}(S) \equiv 0. \tag{2.39}$$

But $\gamma \in M_1 = i^{-1}(\underline{0})$ and (2.36) implies

$$\int_I \gamma_{13}(S)dS = 0. \tag{2.40}$$

The Mean Value Theorem implies there is an $S_0 \in I$ for which $\gamma_{13}(S_0) = 0$. This result and (2.39) imply $\det\gamma(S_0) = 0$, contradicting $\gamma(S_0) \in O(E^3)$. Thus, $\underline{0} \in \mathbb{R}^2$ is a regular value for i, and M_1 is a smooth, codimension 2 submanifold. Finally, (2.37) and $M_1 = i^{-1}(\underline{0})$ imply (2.34). Equation (2.35) follows in a similar manner. ∎

Remark. The isomorphism L_γ of Proposition 2.20 does not restrict to an isomorphism between $T_{\gamma_0}\mathcal{M}_1$ and $T_\gamma\mathcal{M}_1$. So we cannot simply restrict Definition 2.21 to $\mathcal{M}_1$ to obtain a definition of the convective representation and the representation space for those infinitesimal displacements which maintain the constraints. We will resolve this complication in Section IV.

The group Π leaves neither M_1 nor $\mathcal{M}_1$ invariant. We close the subsection by identifying a subgroup which does.

2.25 Lemma. *Let $\mathcal{T}_g$ be given by (2.9). Let*

$$\mathcal{G} = \{\, g \in \Pi \mid \mathcal{T}_g\gamma \in \mathcal{M}_1 \ \forall\ \gamma \in \mathcal{M}_1 \,\}. \tag{2.41}$$

Then

$$\mathcal{G} = G_s \times \Gamma, \tag{2.42}$$

an external direct product, where G_s and Γ are given by Definition 2.4.

Proof. Let $g = (Q_1,Q_2) \in \Pi$ and $\gamma \in \mathcal{M}_1$. By (2.8), (2.9), and the definitions of $\mathcal{M}_1$ and Γ, for $\alpha = 1, 2$, and $\varepsilon = \pm 1$,

$$\int_I \mathcal{T}_g \gamma dS : \underline{e}_\alpha \otimes \underline{e}_3 = \varepsilon \left(x(1) \cdot \underline{e}_3 \right) \left(Q_1^T \underline{e}_\alpha \right) \cdot \underline{e}_3 \ .$$

Choosing $\gamma \in \mathcal{M}_1$ such that $x(1) \cdot \underline{e}_3 \neq 0$, implies $Q_1 \in G_s$, and (2.42) in consequence. ∎

Remark. We examine only Π because we will be analyzing our problem of interest in the convective representation. Conclusions analogous to those of Lemma 2.25 hold for Π_s.

III. THE SPACES OF LOADS

In this section we specify the spaces of dead loads for the rod in the special Cosserat and Kirchhoff theories. As in Section II for the two rod theories we obtain spatial and convective representations for the loads in a given configuration. We determine how the rod loads and their representations transform under the action of the group Π_1. Finally, we generalize the differentiability class for the loads.

III.1 *Loads in the Special Cosserat Theory*

In the special Cosserat theory we can represent the load applied to a rod in a given configuration in terms of a distribution of forces and moments defined along the centerline of the rod [6]. Hence, we may represent them in terms of vector- and anti-symmetric matrix-valued functions defined on the unit interval and at its ends (see Definition 3.7). For rod problems either we may take such a representation as given *a-priori*, or as arising from a three-dimensional force distribution by viewing the rod in its deformed configuration as a three-dimensional body with planar cross sections and suitably "averaging" the force distribution over the sections. It is this second approach we wish to use here.

In general, this latter approach is intractable, because the three-dimensional force distributions change when we change the configuration of the rod. However, if we restrict our attention to *three-dimensional spatially dead* loads (Assumption 3.2), we can carry out the procedure, and in fact, it simplifies. For such force distributions we have the luxury of first averaging each over a single "reference" configuration (the straight configuration for the rod) to produce a space of vector- and matrix-valued functions defined on the unit interval and at its ends (Definition 3.4). It is this space of functions which we call the space of dead rod loads for the special Cosserat theory. Given such a dead rod load, given the rod configuration, we obtain the forces and moments which the load and configuration produce along the centerline by taking the anti-symmetric part of specific matrix products (Definition 3.7). In this manner we obtain a convective and a spatial representation for the rod load. In rigid body mechanics we represent the effects of a force distribution on a rotating body in terms of a torque it produces relative to the body axis or spatial axis. The convective and spatial representations for the dead rod load presented here extend this characterization for the force distribution to the rod models.

In this subsection we present in detail the averaging process, the intermediate space of dead rod loads, and the load representation space for the special Cosserat theory. We develop the spatial and convective representations for a dead rod load in a given configuration, and relate them to the usual descriptions in terms of coordinate functions relative to the spatial or director vector functions. We close the subsection by presenting several examples in which we compute the dead rod load associated with a three-dimensional force distribution, and the subsequent convective representations of the loads produced in various rod configurations.

Remark. This approach to characterizing rod loads was developed in [11] and [12], and later formalized in [6]. Its significance is now being appreciated with the infusion of the ideas of symplectic geometry into continuum mechanics (see [33]).

In the special Cosserat theory we can represent the load applied to a rod in a given configuration in an *a-priori* manner by hypothesizing a particular representation for the work done by it on an infinitesimal displacement from a given configuration ([11], [12]).

3.1 Hypothesis. *In the configuration χ the virtual work done by an applied load on an infinitesimal displacement $H_\chi = (h, H_\gamma)$ may be represented as*

$$\left.\begin{aligned}\mathcal{A} = \int_I [f_\chi(S)\cdot h(S) + q_{\alpha_\chi}(S)\cdot H_\gamma(S)\underline{e}_\alpha]dS + t_\chi(1)\cdot h(1) + \\ p_{\alpha_\chi}(1)\cdot H_\gamma(1)\underline{e}_\alpha + t_\chi(0)\cdot h(0) + p_{\alpha_\chi}(0)\cdot H_\gamma(0)\underline{e}_\alpha = \end{aligned}\right\} \quad (3.1)$$

$$\left.\begin{aligned}\int_I [f_\chi(S)\cdot h(S) + \mathbb{Q}_\chi(S):H_\gamma(S)]dS + \\ t_\chi(1)\cdot h(1) + \mathbb{P}_\chi(1):H_\gamma(1) + t_\chi(0)\cdot h(0) + \mathbb{P}_\chi(0):H_\gamma(0),\end{aligned}\right\} \quad (3.2)$$

where the summation over α = 1, 2 is implied, f_χ, q_{α_χ} are known E^3-valued functions on [0,1], $t_\chi(S)$, $p_{\alpha_\chi}(S)$, S = 0, 1 are known vectors in E^3, and

$$\left.\begin{aligned}\mathbb{Q}_\chi(S) = q_{\alpha_\chi}(S) \otimes \underline{e}_\alpha, \\ \mathbb{P}_\chi(S) = p_{\alpha_\chi}(S) \otimes \underline{e}_\alpha.\end{aligned}\right\} \quad (3.3)$$

■

Interpret f_χ and t_χ as the usual applied body force and traction force densities along the centerline of the rod in the configuration χ. Interpret q_{α_χ} and p_{α_χ}, α = 1, 2 as the generalized body and traction force densities acting on the director displacements.

As the subscript χ indicates, the specification of the various densities in (3.1) generally depends upon the configuration. Call a *load system* the specification of a rule or correspondence which associates with each $\chi \in N$ the collection $(f_\chi, q_{\alpha_\chi}, t_\chi, p_{\alpha_\chi})$ of (3.1), or equivalently, the quartet $(f_\chi, \mathbb{Q}_\chi, t_\chi, \mathbb{P}_\chi)$ of (3.2).

In this work we restrict attention to a particular class of load systems. We discuss other classes in Section VIII.

3.2 Assumption. *Given $k \geq 0$ and $\chi = (x,\gamma) \in N$ restrict attention to those load systems for which there are functions*

$$(f,\ \mathbb{Q}) \in C^k(I, E^3 \times L(E^3)), \quad \mathbb{Q}(S)\cdot \underline{e}_3 = 0 \in E^3,$$

and values

$$t(S) \in E^3,\ \mathbb{P}(S) \in L(E^3), \quad \mathbb{P}(S)\cdot \underline{e}_3 = 0 \in E^3,\ S = 0,\ 1$$

for which

$$\left.\begin{aligned} & f_\chi(S) \equiv f(S) \\ & \mathbb{Q}_\chi(S) \equiv \mathbb{Q}(S),\ 0 \leq S \leq 1, \\ & t_\chi(S) \equiv t(S) \\ & \mathbb{P}_\chi(S) \equiv \mathbb{P}(S),\ S = 0,\ 1. \end{aligned}\right\} \qquad (3.4)$$

■

We can show how load systems which satisfy (3.4) arise when we average over material sections of the rod three-dimensional body and surface force distributions which are themselves independent of the three-dimensional configuration of the body. We summarize the description of such force distributions presented in [1]. We then average over the material cross sections of the rod in the given configuration in the manner of [6] to obtain a representation for the load which satisfies (3.4), and whose virtual work is of the form (3.2).

As in section II.1 view a rod as a slender three-dimensional body $\mathcal{B}$ which we identify with its position in a reference configuration as a circularly cylindrical solid of length one and radius R. Let $\partial\mathcal{B}$ denote the surface of $\mathcal{B}$, and let $\partial\mathcal{B}(S)$ denote the surface of a material section,

$$\partial\mathcal{B}(S) = \left\{ X = (X^1,\ X^2,\ S) \mid (X^1)^2 + (X^2)^2 = R^2 \right\},\ 0 < S < 1.$$

In the three-dimensional theory, the space of C^k configurations for the rod with the left end of its centerline fixed at the origin is

$$\mathfrak{C} = \{ \mathfrak{x} \in C^k(\mathcal{B},E^3) \mid \mathfrak{x}(0,0,0) = 0 \in E^3,\ \mathfrak{x} \text{ is an embedding } \}.$$

As $\mathfrak{C}$ is an open set in a Banach space, the space of C^k three-dimensional infinitesimal displacements at $\mathfrak{x}$, or the tangent space to $\mathfrak{C}$ at $\mathfrak{x}$, is

$$T_{\mathfrak{x}}\mathfrak{C} = \{ V \in C^k(\mathcal{B},E^3) \mid V(0,0,0) = 0 \in E^3 \}.$$

Equation (2.1) relates the configuration space for the special Cosserat theory to the space of three-dimensional configurations. For $\chi = (x,\gamma) \in N$, define $\Lambda(\chi) \in \mathfrak{C}$ by

$$\Lambda(\chi)(X^1,X^2,S) = x(S)+ \phi^{\alpha}(X^1,X^2,S)\gamma(S)\underline{e}_{\alpha} \,. \tag{3.5}$$

Equation (3.5) induces a relation between infinitesimal displacements in the two theories. If $\chi = (x,\gamma) \in N$, then

$$\begin{aligned} T\Lambda(\chi):\ T_{\chi}N &\longrightarrow T_{\Lambda(\chi)}\mathfrak{C} \\ (h,H_{\gamma}) &\longmapsto [T\Lambda(\chi)](h,H_{\gamma}) \equiv V, \end{aligned}$$

where
$$V(X^1,X^2,S) = h(S) + \phi^{\alpha}(X^1,X^2,S)H_{\gamma}(S)\underline{e}_{\alpha}. \tag{3.6}$$

As in [1] let $b\colon \mathcal{B} \longrightarrow E^3$ and $\tau\colon \partial\mathcal{B} \longrightarrow E^3$ denote a given body force density and surface traction density, respectively. Since the densities depend only on the body point in $\mathcal{B}$, (b,τ) represents a system of three-dimensional force distributions which is independent of the configuration of the body. Thus, each such pair represents a system of *dead* or *spatial* loads. For k_1 and k_2 to be specified later, set

$$\mathfrak{L} = \left\{ l_{3D} = (b,\tau) \mid b \in C^{k_1}(\mathcal{B},E^3),\ \tau \in C^{k_2}(\partial\mathcal{B},E^3),\ \int_{\mathcal{B}} b\,dV + \int_{\partial\mathcal{B}} \tau\,dA = 0 \right\}$$

where dV and dA are the volume and area elements on $\mathcal{B}$ and $\partial\mathcal{B}$, respectively, in the reference configuration. The work done by l_{3D} in the configuration $\mathfrak{X} \in \mathfrak{C}$ on an infinitesimal displacement $V \in T_{\mathfrak{X}}\mathfrak{C}$ is given by

$$<l_{3D},V> = \int_{\mathcal{B}} b(X)\cdot V(X)dV + \int_{\partial\mathcal{B}} \tau(X)\cdot V(X)dA. \tag{3.7}$$

The astatic load $k(l_{3D},\mathfrak{X}) \in L(E^3)$ is

$$k(l_{3D},\mathfrak{X}) = \int_{\mathcal{B}} b(X) \otimes \mathfrak{X}(X)dV + \int_{\partial\mathcal{B}} \tau(X) \otimes \mathfrak{X}(X)dA. \tag{3.8}$$

The antisymmetric component of the astatic load specifies the torque the load l_{3D} exerts in the configuration $\mathfrak{X}$, using the relation (2.13).

In the manner of [6] constrain the three-dimensional configurations and the infinitesimal displacements to be those which arise from the special Cosserat theory by means of (3.5). We then obtain a representation for the loads, the astatic load, and the virtual work form for the rod theory which are consistent with Hypothesis 3.1 and which fulfill the requirements of Assumption 3.2.

3.3 Proposition. *Let $\chi = (x,\gamma) \in N$, let $H_\chi = (h,H_\gamma) \in T_\gamma N$, and let $\mathfrak{X} = \Lambda(\chi)$ be given by (3.5). Let $l_{3D} = (b,\tau) \in \mathfrak{L}$.*

a) Then

$$\left.\begin{aligned} <l_{3D},[T\Lambda(\chi)H_\chi> &= \int_I [f(S)\cdot h(S)dS + \mathbb{Q}(S):H_\gamma(S)]dS + \\ t(1)\cdot h(1) + \mathbb{P}(1)&:H_\gamma(1) + t(0)\cdot h(0) + \mathbb{P}(0):H_\gamma(0), \end{aligned}\right\} \tag{3.9}$$

where, for $0 \le S \le 1$, and for $X = (X^1,X^2,S)$,

$$\left.\begin{aligned} f(S) &= \int_{\mathcal{B}(S)} b(X)dA(S) + \int_{\partial\mathcal{B}(S)} \tau(X)dc(S), \\ q^{\alpha}(S) &= \int_{\mathcal{B}(S)} \phi^{\alpha}(X)b(X)dA(S) + \int_{\partial\mathcal{B}(S)} \phi^{\alpha}(X)\tau(X)dc(S), \\ \mathbb{Q}(S) &= q^{\alpha}(S) \otimes \underline{e}_{\alpha}. \end{aligned}\right\} \quad (3.10)$$

For $S = 0, 1$,

$$\left.\begin{aligned} t(S) &= \int_{\mathcal{B}(S)} \tau(X)dA(S), \\ p^{\alpha}(S) &= \int_{\mathcal{B}(S)} \phi^{\alpha}(X)\tau(X)dA(S), \\ \mathbb{P}(S) &= p^{\alpha}(S) \otimes \underline{e}_{\alpha}. \end{aligned}\right\} \quad (3.11)$$

Here, $dA(S)$ is the area element on $\mathcal{B}(S)$, and $dc(S)$ is the induced line element on $\partial\mathcal{B}(S)$.

b) Moreover,

$$\left.\begin{aligned} k(1_{3D},\Lambda(\chi)) = \int_I [f(S) \otimes x(S) + \mathbb{Q}(S)\gamma^{T}(S)]dS + \\ t(1) \otimes x(1) + \mathbb{P}(1)\gamma^{T}(1) + t(0) \otimes \mathbb{P}(0)\gamma^{T}(0). \end{aligned}\right\} \quad (3.12)$$

Proof.

a) By (3.6) and (3.7),

$$\langle 1_{3D}, [T\Lambda(\chi)]H_{\chi}\rangle =$$

$$\int_{\mathcal{B}} b(X)\cdot(h(S) + \phi^{\alpha}(X)H_{\gamma}(S)\underline{e}_{\alpha})dV + \int_{\partial\mathcal{B}} \tau(X)\cdot(h(S) + \phi^{\alpha}(X)H_{\gamma}(S)\underline{e}_{\alpha})dA.$$

In the reference configuration at $X = (X^1,X^2,S)$ in $\mathcal{B}$, $dV = dA(S)dS$, where $dA(S)$ is the induced area element on the material cross section $\mathcal{B}(S)$ on which X resides. Likewise if $X \in \partial\mathcal{B}$, if $0 < S < 1$, $dA(S) = dc(S)dS$, where $dc(S)$ is the line element on $\partial\mathcal{B}(S)$ induced from $\mathcal{B}(S)$. If $S = 0$ or 1, $dA(S)$ is the induced area element on $\mathcal{B}(S) \cap \partial\mathcal{B}$. Expressing the integrals over

$\mathcal{B}$ and $\partial\mathcal{B}$ as iterated integrals over the material cross sections and I, we obtain (3.9), subject to the definitions (3.10) and (3.11).

b) By (3.6) and (3.8),

$$k(1_{3D},\Lambda(\chi)) =$$

$$\int_{\mathcal{B}} b(X) \otimes (x(S) + \phi^{\alpha}(X)\gamma(S)\underline{e}_{\alpha})dV + \int_{\partial\mathcal{B}} \tau(X) \otimes (x(S) + \phi^{\alpha}(X)\gamma(S)\underline{e}_{\alpha})dA.$$

Expressing the integrals over $\mathcal{B}$ and $\partial\mathcal{B}$ as iterated integrals in the manner of part a), we obtain (3.12). ∎

Remarks.

1. We can give a geometric formulation to the perspective of [6], which views the special Cosserat theory for the rod as a constrained three-dimensional theory using fiber bundles. See [15].

2. Assumption 3.2 is an example of the specification of a system of forces in the sense of [16] in the context of the theory of rods.

Equations (3.10) and (3.11) average the three-dimensional spatially-dead force distributions over the reference (straight) configuration for the rod to produce vector and matrix-valued functions defined on the unit interval and at its ends. Given these functions, Proposition 3.3 asserts that the virtual work and the astatic load they produce in any configuration on any virtual displacement can be computed from (3.9) and (3.12). Thus the proposition provides the following representation for the loads, the virtual work form, and the astatic load map for the special Cosserat rod theory and the class of load systems we consider in this work.

3.4 Definition.

a) Given $k \geq 0$, define

$$L = \left\{ \begin{array}{c} l = (f,\mathbb{Q},t,\mathbb{P}) \mid (f,\mathbb{Q} \in C^k(I,E^3 \times L(E^3)),\ \mathbb{Q}(S)\underline{e}_3 = 0, \\ t(S) \in E^3,\ \mathbb{P}(S) \in L(E^3),\ \mathbb{P}(S)\underline{e}_3 = 0,\ S = 0,\ 1 \\ \int_I f(S)dS + t(1) + t(0) = 0 \end{array} \right\}$$

Call L the space of C^k dead loads for the special Cosserat theory in the material representation.

b) For $l = (f,\mathbb{Q},t,\mathbb{P}) \in L$, for $\chi = (x,\gamma) \in N$, let $\Omega(l,\chi)$ be the linear functional on $T_\chi N$ given by

$$\left. \begin{array}{l} \langle\Omega(l,\chi),(h,H_\gamma)\rangle = \\ \qquad \int_I [f(S)\cdot h(S) + \mathbb{Q}(S):H_\gamma(S)]dS + \\ t(1)\cdot h(1) + \mathbb{P}(1):H_\gamma(1) + t(0)\cdot h(0) + \mathbb{P}(0):H_\gamma(0). \end{array} \right\} \quad (3.13)$$

c) Define the astatic load map for the special Cosserat theory

$$\begin{array}{rcl} k:\ L \times N & \longrightarrow & L(E^3) \\ (l,\chi) & \longmapsto & k(l,\chi) \end{array}$$

$$\left. \begin{array}{l} \text{by} \quad k(l,\chi) = \int_I [f(S) \otimes x(S) + \mathbb{Q}(S)\gamma^T(S)]dS + \\ t(1) \otimes x(1) + \mathbb{P}(1)\gamma^T(1) + t(0) \otimes x(0) + \mathbb{P}(0)\gamma^T(0). \end{array} \right\} \quad (3.14)$$

■

We can relate the description of the applied load as an element of L to its usual description in terms of a distribution of forces and moments along the centerline of the rod in the given configuration. We obtain the relation by expressing the work done by an applied load on an infinitesimal displacement in terms of the spatial and convective representation of that displacement.

3.5 Lemma. *Let $l = (f,\mathbb{Q},t,\mathbb{P}) \in L$, $\chi = (x,\gamma) \in N$, and $(h,H_\gamma) \in T_\chi N$. Let $\delta(S)$ be given by Definition 2.12.*

a) If (b_2,Z_2) is the spatial representation for (h,H_γ), then

$$\left.\begin{aligned}&\langle\Omega(1,\chi),(h,H_{\gamma})\rangle = \\ &\int_I[\delta'(S)f(S)\cdot b_2(S) + \delta'(S)skew(\mathbb{Q}(S)\gamma^T(S)):Z_2(S)]dS + \\ &\quad \delta'(1)t(1)\cdot b_2(1) + \delta'(1)skew(\mathbb{P}(1)\gamma^T(1)):Z_2(1) + \\ &\quad \delta'(0)t(0)\cdot b_2(0) + \delta'(0)skew(\mathbb{P}(0)\gamma^T(0)):Z_2(0).\end{aligned}\right\} \quad (3.15)$$

b) If (b,Z) is the convective representation for (h,H_{γ}), then

$$\left.\begin{aligned}&\langle\Omega(1,\chi),(h,H_{\gamma})\rangle = \\ &\int_I[\gamma^T(S)f(S)\cdot b(S) + skew(\gamma^T(S)\mathbb{Q}(S)):Z(S)]dS + \\ &\quad \gamma^T(1)t(1)\cdot b(1) + skew(\gamma^T(1)\mathbb{P}(1)):Z(1) + \\ &\quad \gamma^T(0)t(0)\cdot b(0) + skew(\gamma^T(0)\mathbb{P}(0)):Z(0).\end{aligned}\right\} \quad (3.16)$$

Proof. If $Q \in O(E^3)$, A, $B \in L(E^3)$, then $A:BQ = AQ^T:B$. Moreover, if $Z \in skew(E^3)$, then $B:Z = skew(B):Z$. These properties, (3.13), and Definition 2.12 imply (3.15). Since $A:QB = Q^TA:B$, the above properties, (3.13) and Definition 2.12 imply (3.16). ∎

Given the hypotheses of Lemma 3.5, define two sets of vector functions $(\underline{z}_2(S), \ell_2(S), \underline{\ell}_2(S), t_2(S), \underline{\rho}_2(S))$, and $(\underline{z}(S), \ell(S), \underline{\ell}(S), t(S), \underline{\rho}(S))$ by

$$\left.\begin{aligned}\hat{\underline{z}}_2(S) &= Z_2(S), \\ \hat{\underline{\ell}}_2(S) &= 2\delta'(S)skew(\mathbb{Q}(S)\gamma^T(S)), \quad 0 \le S \le 1, \\ \ell_2(S) &= \delta'(S)f(S),\end{aligned}\right\} \quad (3.17)$$

$$\left.\begin{aligned}\hat{\underline{\rho}}_2(S) &= 2\delta'(S)skew(\mathbb{P}(S)\gamma^T(S)), \\ t_2(S) &= \delta'(S)t(S),\end{aligned}\quad S = 0,\ 1,\right\} \quad (3.18)$$

and

$$\left.\begin{aligned} \hat{z}(S) &= Z(S), \\ \ell(S) &= \gamma^T(S)f(S), \\ \underline{\hat{\ell}}(S) &= 2\text{skew}(\gamma^T(S)\mathbb{Q}(S)), \quad 0 \le S \le 1, \end{aligned}\right\} \tag{3.19}$$

$$\left.\begin{aligned} t(S) &= \gamma^T(S)t(S), \\ \hat{\underline{p}}(S) &= 2\text{skew}(\gamma^T(S)\mathbb{P}(S)), \quad S = 0,\ 1. \end{aligned}\right\} \tag{3.20}$$

The next lemma relates these vector functions and the representations of the virtual work presented in Lemma 3.5.

3.6 Lemma. *Given $l \in L$, $\chi \in N$, and $H_\chi \in T_\chi N$ as in Lemma 3.5, and the vector functions defined by (3.17) through (3.20), then*

a)

$$\left.\begin{aligned} &\langle\Omega(l,\chi),(h,H_\gamma)\rangle = \int_I [f_2(S)\cdot b_2(S) + \underline{\ell}_2(S)\cdot \underline{z}_2(S)]dS + \\ &t_2(1)\cdot b_2(1) + \underline{p}_2(1)\cdot \underline{z}_2(1) + t_2(0)\cdot b_2(0) + \underline{p}_2(0)\cdot \underline{z}_2(0), \end{aligned}\right\} \tag{3.21}$$

where $(b_2, \hat{\underline{z}}_2)$ is the spatial representation for (h, H_γ).

b)

$$\left.\begin{aligned} &\langle\Omega(l,\chi),(h,H_\gamma)\rangle = \int_I [\ell(S)\cdot b(S) + \underline{\ell}(S)\cdot \underline{z}(S)]dS + \\ &t(1)\cdot h(1) + \underline{p}(1)\cdot \underline{z}(1) + t(0)\cdot h(0) + \underline{p}(0)\cdot \underline{z}(0), \end{aligned}\right\} \tag{3.22}$$

where $(b, \hat{\underline{z}})$ is the convective representation for (h, H_γ).

c) If $\underline{d}_j(S) = \gamma(S)\underline{e}_j$, $j = 1, 2, 3$, then for $0 \le S \le 1$,

$$\left.\begin{aligned} \underline{d}_j(S)\cdot\underline{z}_2(S) &= \underline{e}_j\cdot\underline{z}(S),\\ \underline{d}_j(S)\cdot h(S) &= \underline{e}_j\cdot b(S),\\ \underline{d}_j(S)\cdot f(S) &= \underline{e}_j\cdot \ell(S),\\ \underline{d}_j(S)\cdot\underline{\ell}_2(S) &= \underline{e}_j\cdot\underline{\ell}(S), \end{aligned}\right\} \quad (3.23)$$

and for $S = 0, 1$,

$$\left.\begin{aligned} \underline{d}_j(S)\cdot t(S) &= \underline{e}_j\cdot t(S),\\ \underline{d}_j(S)\cdot\underline{\rho}_2(S) &= \underline{e}_j\cdot\underline{\rho}(S). \end{aligned}\right\} \quad (3.24)$$

Proof. Equation (3.21) follows from (3.15), (3.17), (3.18), and Lemma 2.6. Likewise, (3.22) follows from (3.16), (3.19), (3.20), and Lemma 2.6. By Lemma 2.6 and (2.19),

$$\underline{z}_2(S) = \gamma(S)\underline{z}(S).$$

So for $j = 1, 2, 3$,

$$\underline{d}_j(S)\cdot\underline{z}_2(S) = \gamma(S)\underline{e}_j\cdot\underline{z}_2(S) = \underline{e}_j\cdot\underline{z}(S).$$

By Lemma 2.6, (3.17), and (3.19),

$$\underline{\ell}_2(S) = \gamma(S)\underline{\ell}(S).$$

So
$$\underline{d}_j(S)\cdot\underline{\ell}_2(S) = \gamma(S)\underline{e}_j\cdot\gamma(S)\underline{\ell}(S) = \underline{e}_j\cdot\underline{\ell}(S).$$

The other equations of (3.23) and (3.24) follow similarly. ∎

As indicated in [9], [11], and [12], equation (3.21) is the representation of the virtual work done by a distribution $(f,\underline{\ell}_2,t,\underline{\rho}_2)$ of force and moment densities along and at the end of the centerline of a rod in the configuration χ. Thus, equations (3.17) and (3.18) relate the representation of the rod load as an element of L to its usual description in terms of force and moment densities.

Moreover, as indicated in [11] and [12] it is advantageous to represent the force and moment densities and infinitesimal displacements occurring in (3.21) in terms of their components relative to the frame of director vectors $\underline{d}_j(S) = \gamma(S)\underline{e}_j$. Lemma 3.6 indicates that representing the virtual work in this way is equivalent to representing the virtual work of the load $l \in L$ using the convective representation for the infinitesimal displacement, given by (3.16), with the load expressed in terms of a distribution of forces and moments as (3.22), using the vector quantities (3.19) and (3.20). Equation (3.22) is the intrinsic form for the virtual work presented in [9] in this context.

In the light of [7] and [33], we recognize that representing the virtual work in the manner of Lemma 3.6 constitutes the specification of a spatial and a convective representation for the rod load relative to the configuration. We formalize this observation with the following definition.

3.7 Definition. *Let $k \geq 0$. Let $\Delta(S)$ be given by Definition 2.12. Set*

$$T = \left\{ \begin{array}{c} \Xi = (f,\hat{q},t,\hat{p}) \mid (f,\hat{q}) \in C^k(I,E^3 \times skew(E^3)), \\ (t(S),\hat{p}(S)) \in E^3 \times skew(E^3),\ S = 0,\ 1, \\ \int_I f(S)dS + t(1) + t(0) = 0 \end{array} \right\}.$$

For $(l,\chi) \in L \times N$, define $\beta_2(l,\chi)$, $\beta(l,\chi) \in T$ by

$$\beta_2(l,\chi) = (f, 2skew(\mathbb{Q}\gamma^T), t, 2skew(\mathbb{P}\gamma^T))\Delta', \tag{3.25}$$

$$\beta(l,\chi) = (\gamma^T f, 2skew(\gamma^T\mathbb{Q}), \gamma^T t, 2skew(\gamma^T\mathbb{P})). \tag{3.26}$$

Call $\beta_2(l,\chi)$ the spatial representation *of the load $l \in L$ in the configuration χ, and call $\beta(l,\chi)$ its* convective representation. *Call T the* load representation space *for the given space of loads L and configurations N.* ∎

We relate these quantities to the virtual work by means of the following bilinear form.

3.8 Definition. *Let $k \geq 1$. Let D be the representation space given in Definition 2.12. For $\Xi \in T$ and $(v,Z) \in D$ define $K(\Xi,(v,Z)) \in R$ by*

$$\left.\begin{array}{l} K(\Xi,(v,Z)) = \int_I [f(S)\cdot v(S) + (1/2)\hat{q}(S):Z(S)]dS + \\ t(1)\cdot v(1) + (1/2)\hat{p}(1):Z(1) + t(0)\cdot v(0) + (1/2)\hat{p}(0):Z(0). \end{array}\right\} \quad (3.27)$$

■

3.9 Lemma.

a) K is a nondegenerate bilinear pairing.

b) For $l \in L$, $\chi \in N$, and $H_\chi = (h,H_\gamma) \in T_\chi N$,

$$\langle\Omega(l,\chi),H_\chi\rangle = K(\beta_2(l,\chi),(b_2,Z_2)) = K(\beta(l,\chi),(b,Z)), \quad (3.28)$$

where (b_2,Z_2) and (b,Z) are the spatial and convective representations of H_χ, respectively.

Proof.

a) Equation (3.27) indicates that K is a bilinear form. Suppose $\Xi = (f,\hat{q},t,\hat{p}) \in T$ is in $\ker_1 K$. That is to say, $K(\Xi,(b,Z)) = 0$ for all $(b,Z) \in D$. Suppose $\hat{q}(S_0) \neq 0$ for some $S_0 \in I$. Then $\hat{q}(S):\hat{q}(S_0) > 0$ on some neighborhood U of S_0. Choose $b(S) \equiv 0$ and $Z(S) = \zeta(S)\hat{q}(S_0)$, for $\zeta(S)$ a Urysohn function subordinate to U. Then

$$K(\Xi,(b,Z)) = (1/2) \int_U \zeta(S)\hat{q}(S):\hat{q}(S_0)dS > 0,$$

a contradiction. So $\hat{q}(S) \equiv 0$. In a similar manner we obtain $f(S) \equiv 0$. By further suitable choices of (b,Z), we obtain

$$\hat{p}(1):A = 0 \quad \text{and} \quad \hat{p}(0):A = 0$$

for all $A \in \text{skew}(E^3)$, and

$$t(1)\cdot v = 0 \quad \text{and} \quad t(0)\cdot v = 0$$

for all $v \in E^3$. So $\Xi = (0,0,0,0)$, implying $\ker_1 K = 0$. Likewise, assuming $K(\Xi,(b,Z)) = 0$ for all $\Xi \in T$, suitable choices for Ξ imply $(b,Z) \equiv 0$. Thus, $\ker_2 K = 0$. As both kernels are trivial, K is a nondegenerate bilinear form.

b) The first equality follows from (3.15), (3.25), and (3.27). The second equality follows from (3.16), (3.26), and (3.28). ∎

As shown by J. Simo, J. Marsden, and P. Krishnaprasad in [33], the spatial and convective representations (3.25) and (3.26) generalize to the rod theory the representations we obtain in rigid body mechanics when we characterize the effects of a force distribution on a rotating body in terms of a torque relative to the spatial or to body axis.

Definition 3.7 Lemma 3.9 formalize how a load and a configuration determine the force and moment distributions which characterize the ability of the load to do work on an infinitesimal displacement from the configuration. As (3.25) and (3.26) indicate, representing the load as an element of T involves the configuration. The significance of this involvement will become clear in Section III.3.

We close this subsection by presenting some examples of three-dimensional spatial loads which we average to produce loads in L. We then represent them in various configurations as force and moment distributions in terms of elements of T.

3.10 Example. *An Axially Symmetric, Symmetric Compressive Load.* Given $p > 0$, take $l_{3D} = (0,\tau)$, where for $X = (X^1,X^2,S) \in \partial\mathcal{B}$,

$$\tau(X) = \begin{cases} 0 & 0 < S < 1 \\ \dfrac{p}{\pi R^2}\,\underline{e}_3 & S = 0 \\ \dfrac{-p}{\pi R^2}\,\underline{e}_3 & S = 1 \end{cases} .$$

This surface distribution corresponds to an axially symmetric end compression. Averaging over the cross sections using (3.10) and (3.11), taking $\phi^{\alpha}(X) = X^{\alpha}$ gives the following characterization of the load as an element of L:

$$l = (f, \mathbb{Q}, t, \mathbb{P}) = (0,0,t,0) \in L,$$

where
$$t(0) = -t(1) = p\underset{=}{e}_3.$$

Let $\chi_0 = (x_0, \gamma_0)$ denote the reference configuration (see Theorem 2.9). Let $Q_\phi \in O(2)$ be a counterclockwise rotation through the angle ϕ about the $\underset{=}{e}_3$ axis. Let $g_\phi = (Q_\phi, I) \in \Pi$. Then by (3.25) the spatial representation of the load l in the configuration $g_\phi \chi_0$ is

$$\beta_2(l, g_\phi \chi_0) = (0,0,t,0).$$

The result indicates that l produces no distribution of moments about the centerline in the configuration $g_\phi \chi_0$, as expected.

3.11 Example. *A Parallel Force System.*
Take $l_{3D} = (0, \tau)$, where for $X = (X^1, X^2, S) \in \partial\mathcal{B}$,

$$\tau(X) = \begin{cases} X^1 \underset{=}{e}_1 & 0 < S < 1 \\ 0 & S = 0,\ 1 \end{cases} .$$

Such a uni-directional force distribution is an example of a three-dimensional parallel load system ([3], p. 207). Averaging over the cross section as in the previous example gives the characterization of l_{3D} as an element of L:

$$l = (f, \mathbb{Q}, t, \mathbb{P}) = (0, \mathbb{Q}, 0, 0),$$

where
$$\mathbb{Q}(S) = \pi R^3 \underset{=}{e}_1 \otimes \underset{=}{e}_1.$$

Let $g_\phi \chi_0$ represent the rotated reference configuration defined in Example 3.10. Then

$$\beta_2(1, g_\phi \chi_0) = (0, \hat{\underline{\ell}}_2, 0, 0),$$

$$\hat{\underline{\ell}}_2(S) = 2\text{skew}(\mathbb{Q}(S) Q_\phi^T) = 2\text{skew}(\pi R^3 \underline{e}_1 \otimes Q_\phi \underline{e}_1) = -(\pi R^3)\hat{\underline{e}}_3 \sin\phi,$$

using $Q_\phi \underline{e}_1 = \cos\phi \underline{e}_1 + \sin\phi\ \underline{e}_2$ and Lemma 2.6. As expected, $\underline{\ell}_2(S)$ is the moment density l_{3D} exerts about the centerline in the configuration $g_\phi \chi_0$.

Figure 3.1.1 depicts a cross section of the rod in four configurations $g_\phi \chi_0$, where $\phi = 0, \pi/4, 3\pi/4, \pi$. Superimposed is a vector field depicting the three-dimensional spatial surface force distribution τ on the cross section. For each configuration we indicate the spatial representation of the load, or moment about the centerline.

3.12 Example.

a) A Non-Parallel Force Distribution.

Take $l_{3D} = (0, \tau)$, where for $X = (X^1, X^2, S) \in \partial\mathcal{B}$,

$$\tau(X) = \begin{cases} X^1 \underline{e}_1 - X^2 \underline{e}_2 & 0 < S < 1 \\ 0 & S = 0, 1 \end{cases}.$$

Averaging over the cross section in the reference configuration gives

$$l = (0, \mathbb{Q}, 0, 0),$$

$$\mathbb{Q}(S) = \pi R^3 [\underline{e}_1 \otimes \underline{e}_1 - \underline{e}_2 \otimes \underline{e}_2].$$

Let $g_\phi \chi_0$ represent the rotated reference configurations as in Example 3.10. Then

$$\beta_2(1, g_\phi \chi_0) = (0, 0, 0, 0),$$

since

$$\text{skew}(\underline{e}_1 \otimes Q_\phi \underline{e}_1 - \underline{e}_2 \otimes Q_\phi \underline{e}_2) = (\sin\phi)\text{skew}(\underline{e}_1 \otimes \underline{e}_2 + \underline{e}_2 \otimes \underline{e}_1) = 0.$$

b) A Radial Force Distribution

In contrast, take $l_{3D} = (0,\tau)$, where for $X = (X^1, X^2, S) \in \partial\mathcal{B}$,

$$\tau(X) = \begin{cases} X^1\underline{e}_1 + X^2\underline{e}_2 & 0 < S < 1 \\ 0 & S = 0,\ 1 \end{cases} .$$

Averaging over the reference configuration gives

$$l = (0, \mathbb{Q}, 0, 0),$$
$$\mathbb{Q}(S) = \pi R^3[\underline{e}_1 \otimes \underline{e}_1 + \underline{e}_2 \otimes \underline{e}_2].$$

Let $g_\phi\chi_0$ represents a rotation of the reference configuration, then

$$\beta_2(l, g_\phi\chi_0) = (0, \underline{\ell}_2, 0, 0),$$

for $\hat{\underline{\ell}}_2 = -2\sin\phi\ \hat{\underline{e}}_3$, since

$$\text{skew}(\underline{e}_1 \otimes Q_\phi\underline{e}_1 + \underline{e}_2 \otimes Q_\phi\underline{e}_2) = (\sin\phi)\text{skew}(\underline{e}_1 \otimes \underline{e}_2 - \underline{e}_2 \otimes \underline{e}_1) = 0.$$

Figures 3.1.2 and 3.1.3 depict the two force distributions on a cross section of the rod in the four configurations introduced in Example 3.10. Computing the moments about the centerline directly for each configuration confirms the above results for $\beta_2(l, g_\phi\chi_0)$.

3.13 Example. *A Constant Gravitational Field.*

Take $l_{3D} = (b, 0)$, for

$$b(X) \equiv -\underline{e}_2.$$

Averaging over the cross section gives

$$l = (f, 0, 0, 0),$$

for

$$f(S) \equiv -\pi R^2\underline{e}_2.$$

For $g_\phi \chi_0$ a rotation of the reference configuration,

$$\beta_2(l, g_\phi \chi_0) = (f, 0, 0, 0),$$

indicating that l produces no moment about the centerline in this configuration, as expected. See Figure 3.1.4.

Examples of Admissible Rod Loads

Example 3.11

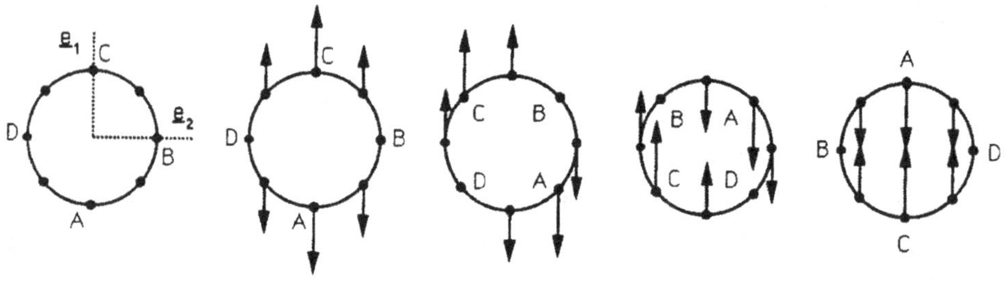

reference	χ_0	$g_1\chi_0$	$g_2\chi_0$	$g_3\chi_0$
$\underline{\ell}_2$	$= \underline{0}$	$\neq \underline{0}$	$\neq \underline{0}$	$= \underline{0}$

Figure 3.1.1

Examples of Admissible Rod Loads (Continued)

Example 3.12

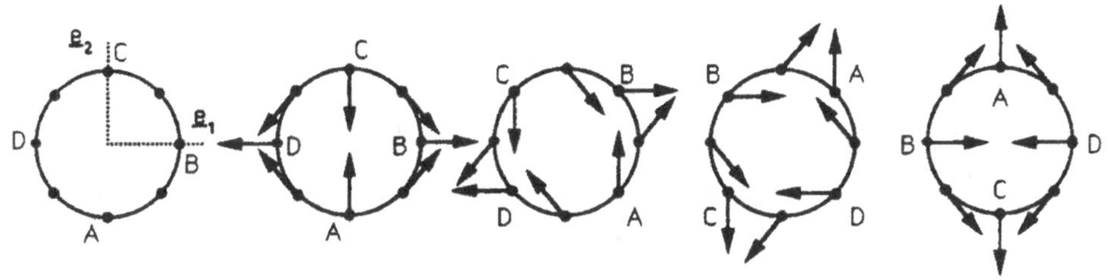

reference	χ_0	$g_1\chi_0$	$g_2\chi_0$	$g_3\chi_0$
$\underline{\ell}_2$	$= \underline{0}$	$= \underline{0}$	$= \underline{0}$	$= \underline{0}$

Figure 3.1.2

Example 3.13

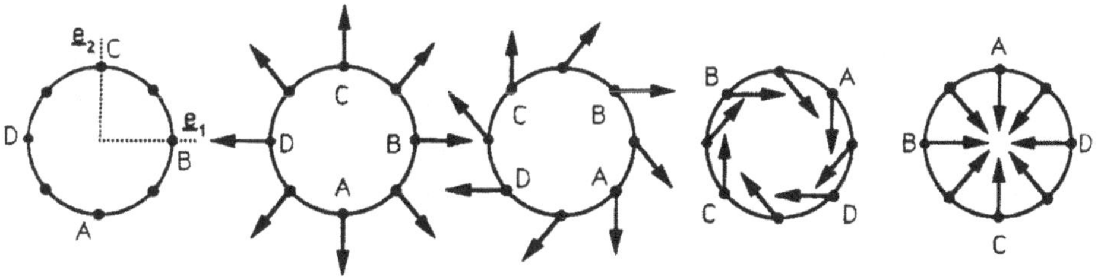

reference	χ_0	$g_1\chi_0$	$g_2\chi_0$	$g_3\chi_0$
$\underline{\ell}_2$	$= \underline{0}$	$\neq \underline{0}$	$\neq \underline{0}$	$= \underline{0}$

Figure 3.1.3

Examples of Admissible Rod Loads (Continued)

Example 3.14

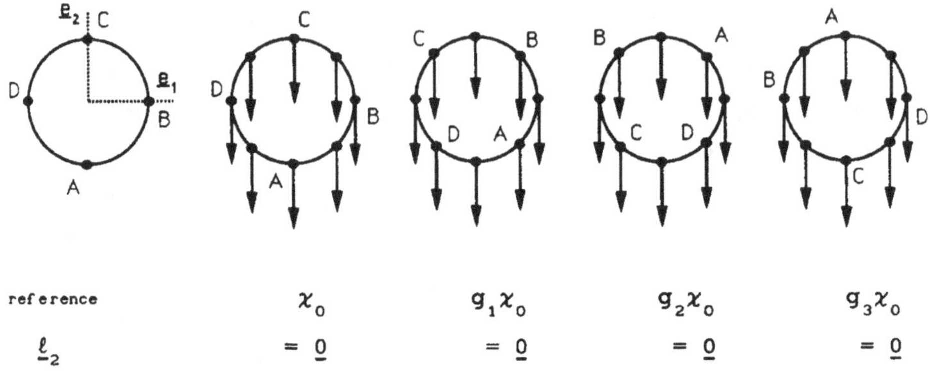

Figure 3.1.4

III.2. *The Space of Loads for the Kirchhoff Theory*

When we restrict attention to those deformations for which the Kirchhoff hypothesis (2.5) is satisfied, the characterization of the load representation space, the dead rod loads, their convective and spatial representations, and the representation of the virtual work presented in the previous subsection simplify.

3.14 Lemma. *Let $\gamma \in M$ and $H_\gamma \in T_\gamma M$. Let $\chi = (x,\gamma) \in N$ and $H_\chi = (h,H_\gamma) \in T_\chi N$ be the corresponding configurations in the special Cosserat theory, as given by Proposition 2.9. Let $l = (f,\mathbb{Q},t,\mathbb{P}) \in L$ be a load in the special Cosserat theory.*

a) Then

$$\left.\begin{aligned}
&\langle\Omega(1,\chi),H_\chi\rangle = \int_I \mu(S):H_\gamma(S)dS + \nu(1):H_\gamma(1) + \nu(0):H_\gamma(0),\\
&\mu(S) = \mathbb{Q}(S) + \left(\int_S^1 f(r)dr + t(1)\right)\otimes \underline{e}_3,\\
&\nu(S) = \mathbb{P}(S), \quad S = 0,\ 1.
\end{aligned}\right\} \quad (3.29)$$

b) The astatic load is

$$k(1,\chi) = \int_I \mu(S)\gamma^T(S)dS + \nu(1)\gamma^T(1) + \nu(0)\gamma^T(0).$$

c) If $Z(S)$ and $Z_2(S)$ are the convective and spatial representations of H_γ, respectively, then

$$\langle\Omega(1,\chi),H_\chi\rangle = \int_I \mathrm{skew}(\gamma^T(S)\nu(S)):Z(S)dS +$$

$$\mathrm{skew}(\gamma^T(1)\nu(1)):Z(1) + \mathrm{skew}(\gamma^T(0)\nu(0)):Z(0) =$$

$$\int_I \Delta'(S)\mathrm{skew}(\mu(S)\gamma^T(S)):Z_2(S)dS +$$

$$\Delta'(1)\mathrm{skew}(\mu(1)\gamma^T(1)):Z_2(1) + \Delta'(0)\mathrm{skew}(\mu(0)\gamma^T(0)):Z_2(0).$$

Proof. Since

$$\frac{d}{dS}\left(\int_S^1 f(r)dr\right) = -f(S),$$

integrating (3.13) by parts and using the definitions of μ and ν gives a). Integrating (3.14) by parts and using (3.29) gives b). Part c) follows from b) in the manner of the proof of Lemma 3.5. ∎

As a consequence of Lemma 3.14, Definition 3.4 for the load space, the virtual work, and the astatic load simplify for the Kirchhoff theory.

3.15 Definition. *Let $k \geq 0$.*

a) Define

$$J = \left\{ \begin{array}{c} \xi = (\mu,\nu) \mid \xi \in C^k(I,L(E^3)),\ \nu(S) \in L(E^3), \\ \nu(S)\underline{e}_3 = 0,\ S = 0,\ 1 \end{array} \right\} \tag{3.30}$$

Call J the space of C^k dead loads *for the Kirchhoff theory in* the material representation.

b) For $\gamma \in M$, $\xi = (\mu,\nu) \in J$ define the virtual work form associated with the load and the configuration to be the linear functional $\Omega(\xi,\gamma)$ on $T_\gamma M$ given by

$$\langle\Omega(\xi,\gamma),H_\gamma\rangle = \int_I \mu(S):H_\gamma(S)dS + \nu(1):H_\gamma(1) + \nu(0):H_\gamma(0). \tag{3.31}$$

c) *Define the astatic load for the Kirchhoff theory to be*

$$k(\xi,\gamma) = \int_I \mu(S)\gamma^T(S)dS + \nu(1)\gamma^T(1) + \nu(0)\gamma^T(0). \tag{3.32}$$

∎

Similarly, Lemma 3.14 simplifies the convective and spatial representations for the load and the virtual work for the Kirchhoff theory which arises from Definition 3.7, 3.8, and Lemma 3.9.

3.16 Definition. *Let $k \geq 0$. Let $\Delta(S)$ be given by Definition 2.12.*

a) Set

$$F = \left\{ \begin{array}{c} \zeta = (\hat{q},\hat{p}) \mid \hat{q} \in C^k(I,skew(E^3)), \\ \hat{p}(S) \in skew(E^3),\ S = 0,\ 1 \end{array} \right\} \tag{3.33}$$

For $\xi = (\mu,\nu)$ and $\gamma \in M$ define $\beta_2(\xi,\gamma)$ and $\beta(\xi,\gamma) \in F$ by

$$\beta_2(\xi,\gamma) = (2skew(\mu\gamma^T), 2skew(\nu\gamma^T))\Delta', \tag{3.34}$$

$$\beta(\xi,\gamma) = (2skew(\gamma^T\mu), 2skew(\gamma^T\nu)). \tag{3.35}$$

b) Let C be the representation space given in Definition 2.12. for $\zeta = (\hat{q},\hat{p}) \in F$ and $Z \in C$, define $K(\zeta,Z) \in \mathbb{R}$ by

$$\left.\begin{aligned} K(\zeta,Z) = (1/2)\int_I \hat{q}(S):Z(S)dS + \\ (1/2)\hat{p}(1):Z(1) + (1/2)\hat{p}(0):Z(0). \end{aligned}\right\} \tag{3.36}$$

■

3.17 Lemma.

a) K is an non-degenerate bilinear pairing.

b) For $\xi \in J$, $\gamma \in M$, $H_\gamma \in T_\gamma M$,

$$\langle\Omega(\xi,\gamma),H_\gamma\rangle = K(\beta_2(\xi,\gamma),Z_2) = K(\beta(\xi,\gamma),Z), \tag{3.37}$$

where Z_2 and Z are the spatial and convective representations for H_γ, respectively.

Proof. Part a) follows in the manner of Lemma 3.9. Part b) follows from Lemma 3.9 by using Lemma 3.14 and Definition 3.16. ■

In the light of Lemma 3.17 and [17], call $\beta_2(\xi,\gamma)$ and $\beta(\xi,\gamma)$ the *spatial* and *convective representations*, respectively, for the load ξ in the configuration γ for the Kirchhoff theory. Call F the *representation space* for the given spaces of loads and configurations for the Kirchhoff theory.

Equations (3.34) through (3.37) formalize how a load in J and a configuration in M determine a distribution of moments along the centerline of the rod in the Kirchhoff theory. As we mentioned in the previous section, these representations are the counterparts in the Kirchhoff rod model of the representations we obtain in rigid body mechanics when we characterize the effect of a force distribution on a rotating body in terms of a torque relative to its spatial and body axes. Notice that representing a load in the Kirchhoff theory in this way involves the configuration. The significance of the involvement will become clear in Section III.3.

Remark. The simplifications in the load space, the virtual work form, the representation space, and the spatial and convective representations for the loads we obtained in this section arise because we are viewing the Kirchhoff rod model as a special Cosserat rod model subject to a constraint. From the perspective of symplectic mechanics, our constructions are the result of a very specific process known as a *reduction* of the unconstrained model. For example, the representation space F arises from the representation space T of the previous section under the reduction. Reduction is a very valuable tool in symplectic mechanics, and its extension to continuum mechanical models will doubtlessly prove to be extremely fertile. The interested reader is strongly invited to review [17].

We close this subsection by representing the loads presented in Examples 3.10 through 3.13 as they would appear in the Kirchhoff theory.

3.18 Example. *An Axially Symmetric, Compressive Load.*
By (3.29) the three-dimensional force distribution of Example 3.10 is characterized in the Kirchhoff theory by an element of J as

$$\xi = (\mu,0), \quad \mu(S) = -p(\underline{e}_3 \otimes \underline{e}_3).$$

If $g_\phi\gamma_0$ represents a rotation of the reference configuration as introduced in Example 3.10, then $Q_\phi\underline{e}_3 = \underline{e}_3$ implies

$$\beta_2(\xi,g_\phi\gamma_0) = (2(-p)\text{skew}(\underline{e}_3 \otimes \underline{e}_3),0) = (0,0),$$

$$\beta(\xi,g_\phi\gamma_0) = (2(-p)\text{skew}(Q_\phi^T\underline{e}_3 \otimes \underline{e}_3),0) = (0,0).$$

3.19 Example. *A Parallel Force Distribution.*
For the Kirchhoff theory the description of Example 3.11 simplifies to

$$\xi = (\mu,0), \quad \mu(S) = \pi R^3\underline{e}_1 \otimes \underline{e}_1,$$

$$\beta_2(\xi,g_\phi\gamma_0) = (2\pi R^3 \mathrm{skew}(\underline{e}_1 \otimes Q_\phi \underline{e}_1),0) = (-(\pi R^3 \sin\phi)\hat{\underline{e}}_3,0),$$

$$\beta(\xi,g_\phi\gamma_0) = (2\pi R^3 \mathrm{skew}(Q_\phi^T \underline{e}_1 \otimes \underline{e}_1),0) = (+(\pi R^3 \sin\phi)\hat{\underline{e}}_3,0),$$

using $Q_\phi \underline{e}_1 = \cos\phi \underline{e}_1 + \sin\phi \underline{e}_2$ and Lemma 2.6.

3.20 Example.

a) A Non-Parallel Force Distribution.

For the Kirchhoff theory, the description of Example 3.12 a) simplifies to

$$\xi = (\mu,0), \quad \mu(S) = \pi R^3(\underline{e}_1 \otimes \underline{e}_1 - \underline{e}_2 \otimes \underline{e}_2),$$

$$\beta_2(\xi,g_\phi\gamma_0) = (2\pi R^3 \mathrm{skew}(\underline{e}_1 \otimes Q_\phi \underline{e}_1 - \underline{e}_2 \otimes Q_\phi \underline{e}_2),\ 0) = (0,0),$$

$$\beta(\xi,g_\phi\gamma_0) = (0,0).$$

b) A Radial Force Distribution

The description of Example 3.12 b) simplifies to

$$\xi = (\mu,0), \quad \mu(S) = \pi R^3(\underline{e}_1 \otimes \underline{e}_1 + \underline{e}_2 \otimes \underline{e}_2),$$

$$\beta_2(\xi,g_\phi\gamma_0) = (-2\pi R^3 \hat{\underline{e}}_3 \sin\phi,0)$$

$$\beta(\xi,g_\phi\gamma_0) = (2\pi R^3 \hat{\underline{e}}_3 \sin\phi,0).$$

3.21 Example. *A Constant Gravitational Field.*

For the Kirchhoff theory Example 3.13 simplifies to

$$\xi = (\mu,0), \quad \mu(S) = (-\pi R^2)(1 - S)\underline{e}_2 \otimes \underline{e}_3,$$

$$\beta_2(\xi,g_\phi\gamma_0) = (\pi R^2(1 - S)\hat{\underline{e}}_1,0),$$

$$\beta(\xi,g_\phi\gamma_0) = (-\pi R^2(1 - S)[\cos\phi\ \hat{\underline{e}}_2 - \sin\phi\ \hat{\underline{e}}_1],0),$$

using Lemma 2.6.

III.3. *The Co-Adjoint Group Action on the Spaces of Loads*

In each of the rod theories the group Π_1 introduced in Section II.1 acts as a group of transformations on the space of loads in a manner which is contragredient, or co-adjoint to its action on the space of configurations. We describe this action in this subsection and develop its spatial and convective representations on the load representation spaces.

3.22 Definition. *Let M, J and N, L be the spaces of configurations and loads in the Kirchhoff and special Cosserat theories, respectively. For $g = (Q_1,Q_2) \in \Pi_1$, for $l = (f,\mathbb{Q},t,\mathbb{P}) \in L$ and for $\xi = (\mu,\nu) \in J$, define*

$$gl = (Q_1 f, Q_1 \mathbb{Q} Q_2^T, Q_1 t, Q_1 \mathbb{P} Q_2^T), \tag{3.38}$$

$$g\xi = (Q_1 \mu Q_2^T, Q_1 \nu Q_2^T). \tag{3.39}$$

■

3.23 Lemma. *Let $g \in \Pi_1$, $l \in L$, and $\xi \in J$.*

a) Then Definition 3.22 specifies an action of Π_1 on L and J as a group of transformations. The action is effective on L. When restricted to Π or Π_s, the action is effective on J.

b) If $\chi \in N$ and if $H_\chi = (h, H_\gamma) \in T_\chi N$, then (3.13) and the group action satisfy the equation

$$\langle\Omega(gl, g\chi), g\cdot H_\chi\rangle = \langle\Omega(l,\chi), H_\chi\rangle.$$

c) If $\gamma \in M$ and if $H_\gamma \in T_\gamma M$, then (3.31) and the group action satisfy the equation

$$\langle\Omega(g\xi, g\gamma), g\cdot H_\gamma\rangle = \langle\Omega(\xi,\gamma), H_\gamma\rangle.$$

Here, $g\cdot H_\chi$ and $g\cdot H_\gamma$ are defined in Proposition 2.13.

Proof. We establish part a) by direct computation, in the manner of Lemma 2.5. By Proposition 2.13, (3.13), and Definition 3.22,

$$<\Omega(gl,g\chi),H_\chi> = \int_I [Q_1 f(S)\cdot Q_1 h(S) + Q_1 \mathbb{Q}(S)Q_2^T : Q_1 H_\gamma(S)Q_2^T]dS +$$
$$Q_1 t(1)\cdot Q_1 h(1) + Q_1 \mathbb{P}(1)Q_2^T : Q_1 H_\gamma(1)Q_2^T +$$
$$Q_1 t(0)\cdot Q_1 h(0) + Q_1 \mathbb{P}(0)Q_2^T : Q_1 H_\gamma(0)Q_2^T$$

Part b) now follows from (3.13) and the orthogonal nature of Q_1 and Q_2. Part c) follows from Proposition 3.11, (3.31), and Definition 3.22 in a manner analogous to part b). ∎

Lemma 3.23 indicates that for each rod theory the action of Π_1 on the space of loads is contragredient, or co-adjoint to its action on the space of configurations relative to the virtual work form. The result is important physically. If $g = (Q,I) \in \Pi_1$ for Q a rotation, then the results says that the work done by a load l in the rotated configuration $g\chi$ on a virtual displacement from that configuration is equivalent to the work done by the "oppositely rotated" load $g^T l$ on a corresponding virtual displacement from the original configuration χ.

Remark. The proof of Lemma 3.23 depends significantly on the "dead" nature of the load, as given by the definition of L and (3.13). If the load were built, for example, from three-dimensional force distributions which varied with the configuration ("live" three-dimensional force distributions), then Lemma 3.23 would fail to hold, and Definition 3.22 would not be the group action co-adjoint to the action on the space of configurations relative to the virtual work form.

We now examine how the spatial and convective representations for the load in the special Cosserat theory respond to the action of Π_1 on L and N.

3.24 Lemma. *Let $g = (Q_1, Q_2) \in \Pi_1$, $l = (f,\mathbb{Q},t,\mathbb{P}) \in L$, and $\chi = (x,\gamma) \in N$. Let $\beta_2(l,\chi) = (f_2,\hat{q}_2,t_2,\hat{p}_2)$ and $\beta(l,\chi) = (\mathfrak{f},\hat{q},t,\hat{p})$ be given by Definition 3.7.*

a) Then

$$\beta_2(gl,g\chi) = (Q_1 f_2, Q_1\hat{q}_2 Q_1^T, Q_1 t_2, Q_1\hat{p}_2 Q_1^T),$$

and $\beta(gl,g\chi) = (Q_2\ell, Q_2\hat{q}Q_2^T, Q_2t, Q_2\hat{p}Q_2^T)$.

b) For $(b,Z) \in D = C^k(I,E^3 \times skew(E^3))$,

$$K(\beta_2(gl,g\chi),\mathcal{B}_2(g)(b,Z)) = K(\beta_2(l,\chi),(b,Z)),$$

and $$K(\beta(gl,g\chi),\mathcal{B}(g)(b,Z)) = K(\beta(l,\chi),(b,Z)),$$

for $\mathcal{B}_2(g)$ *and* $\mathcal{B}(g)$ *specified in Definition 2.15 and* K *specified in Definition 3.8.*

Proof. Part a) follows from Definition 3.7 by a direct computation. By Proposition 2.14, Lemmas 3.9 and 3.29, and the orthogonal nature of Q_1,

$$K(\beta(gl,g\chi),\mathcal{B}(g)(b,Z)) = K(\beta(gl,g\chi),(Q_2b,Q_2ZQ_2^T)) =$$

$$\int_I [Q_2\ell\cdot Q_2b + Q_2\hat{q}Q_2:Q_2ZQ_2^T]dS + Q_2t(1)\cdot Q_2b(1) + Q_2\hat{p}(1)Q_2^T:Q_2Z(1)Q_2^T$$
$$+ Q_2t(0)\cdot Q_2b(0) + Q_2\hat{p}(0)Q_2^T:Q_2Z(0)Q_2^T$$
$$= K(\beta(l,\chi),(b,Z)).$$

Thus the convective version of part b) follows. The spatial version of part b) follows similarly. ∎

In consequence, we can define representations of the group Π_1 on the load spaces of Definition 3.7 which are co-adjoint to the spatial and convective representations of the action of Π_1 on the space of infinitesimal displacements.

3.25 Definition. *Let* $g = (Q_1,Q_2) \in \Pi_1$.

a) If $\Xi = (f,\hat{q},t,\hat{p}) \in T$, *define* $\mathcal{A}_2(g)\Xi \in T$ *by*

$$\mathcal{A}_2(g)\Xi = (Q_1f, Q_1\hat{q}Q_1^T, Q_1t, Q_1\hat{p}Q_1^T). \tag{3.40}$$

b) *Define* $\mathcal{A}(g)\Xi \in T$ *by*

$$\mathcal{A}(g)\Xi = (Q_2f, Q_2\hat{q}Q_2^T, Q_2t, Q_2\hat{p}Q_2^T). \tag{3.41}$$

Call $\mathcal{A}_2$ and $\mathcal{A}$ the spatial and convective representation of Π_1 on the load representation space T in the special Cosserat theory. ∎

3.26 Lemma. *Let D be the representation space given in Definition 2.12. If $g \in \Pi_1$, $\Xi \in T$, and $(b,Z) \in D$, then*

$$K(\mathcal{A}_2(g)\Xi,\mathcal{B}_2(g)(b,Z)) = K(\Xi,(b,Z)) = K(\mathcal{A}(g)\Xi,\mathcal{B}(g)(b,Z)), \tag{3.42}$$

where $\mathcal{B}_2(g)$ and $\mathcal{B}(g)$ are given by Definition 2.15.

Proof. The lemma follows from Definition 3.25 and Lemma 3.24. ∎

Likewise, we can develop the spatial and convective representations for Π_e and Π_1 on the load representation space J which is co-adjoint to its representations on the spaces of virtual displacements in the Kirchhoff theory. We obtain results analogous to Lemma 3.24, 3.26, and Definition 3.25. We summarize these results without proof.

3.27 Lemma. *Let $g = (Q_1,Q_2) \in \Pi_1$, $\xi = (\mu,\nu) \in J$, and $\gamma \in N$. Let $\beta_2(\xi,\gamma) = (\hat{q}_2,\hat{p}_2) \in F$ and $\beta(\xi,\gamma) = (\hat{q},\hat{p}) \in F$ be given by Definition 3.16.*

a) Then

$$\beta_2(g\xi,g\gamma) = (Q_1\hat{q}_2Q_1^T,Q_1\hat{p}_2Q_1^T),$$

$$\beta(g\xi,g\gamma) = (Q_2\hat{q}Q_2^T,Q_2\hat{p}Q_2^T).$$

b) Let C be as specified in Definition 2.12. If $Z \in C$, then

$$K(\beta_2(g\xi,g\gamma),\mathcal{B}_2(g)Z) = K(\beta_2(\xi,\gamma),Z),$$

$$K(\beta(g\xi,g\gamma),\mathcal{B}(g)Z) = K(\beta(\xi,\gamma),Z), \tag{3.43}$$

where $\mathcal{B}_2(g)$ and $\mathcal{B}(g)$ are specified in Definition 2.15. ∎

3.28 Definition. *If $g = (Q_1,Q_2) \in \Pi_1$, and if $\zeta = (\hat{q},\hat{p}) \in F$, define $\mathcal{A}_2(g)\zeta \in F$ and $\mathcal{A}(g)\zeta \in F$ by*

$$\mathcal{A}_2(g)\zeta = (Q_1\hat{q}Q_1^T, Q_1\hat{p}Q_1^T),$$

$$\mathcal{A}(g)\zeta = (Q_2\hat{q}Q_2^T, Q_2\hat{p}Q_2^T). \tag{3.44}$$

When restricted to Π (respectively, Π_s) call $\mathcal{A}$ (respectively, $\mathcal{A}_2$) the convective *(respectively,* spatial*)* representation *for Π (respectively, Π_s) on the load representation space J in the Kirchhoff theory.* ∎

Remark. As with Definition 2.15, we restrict the definition to representations of Π and Π_s so as to obtain effective actions on J.

3.29 Lemma. *Let C be as specified in Definition 2.12. Let $g \in \Pi_1$, $\zeta \in F$, and $Z \in C$.*

a) Then

$$K(\mathcal{A}_2(g)\zeta, \mathcal{B}_2(g)Z) = K(\zeta, Z) = K(\mathcal{A}(g), \mathcal{B}(g)Z), \tag{3.45}$$

where $\mathcal{B}_2(g)$ and $\mathcal{B}(g)$ are given by (2.23) and (2.24), respectively.

b) Moreover,

$$\beta_2(g\xi, g\gamma) = \mathcal{A}_2(g)\beta_2(\xi,\gamma),$$

$$\beta(g\xi, g\gamma) = \mathcal{A}(g)\beta(\xi,\gamma) \tag{3.46}$$

Proof. Equation (3.45) follows from (3.36), the definitions of $\mathcal{B}_2$ and $\mathcal{B}$, and Definition 3.28. Part b) follows from Definitions 3.16 and 3.28. ∎

III.4. *The Generalization of the Load Spaces*

In section V we will use the theory of elliptic differential equations to examine an equilibrium problem for a rod in the Kirchhoff theory which is subject to constraints on its ends.

Consequently, we introduce rod loads whose differentiability class is more general than that introduced previously.

3.30 Definition. *Let $k \geq 1$, and let r be an integer satisfying $r > k + 1/2$. Define*

$$\mathcal{L} = \left\{ \begin{array}{l} l = (f,\mathbb{Q},t,\mathbb{P}) \mid (f,\mathbb{Q}) \in W^{r-2,2}(I,E^3 \times L(E^3)),\ \mathbb{Q}(S)\underline{e}_3 = 0, \\ \qquad (t(S),\mathbb{P}(S)) \in E^3 \times L(E^3),\ \mathbb{P}(S)\underline{e}_3 = 0,\ S = 0,\ 1 \end{array} \right\}$$

$$\mathcal{J} = \{\ \xi = (\mu,\nu) \mid \mu \in W^{r-2,2}(I,L(E^3)),\ \nu(S)\underline{e}_3 = 0,\ S = 0,\ 1\ \}$$

$$\mathcal{T} = \left\{ \begin{array}{l} \Xi = (f,\hat{q},t,\hat{p}) \mid (f,\hat{q}) \in W^{r-2,2}(I,E^3 \times skew(E^3)), \\ \qquad (t(S),\hat{p}(S)) \in E^3 \times skew(E^3),\ S = 0,\ 1 \end{array} \right\}$$

$$\mathcal{F} = \left\{ \begin{array}{l} \zeta=(\hat{q},\hat{p}) \mid \hat{q} \in W^{r-2,2}(I,skew(E^3)), \\ \qquad \hat{p}(S) \in skew(E^3),\ S = 0,\ 1 \end{array} \right\}. \quad \blacksquare$$

Here, the differentiability class $W^{r,2}$ is described in Definition 2.17. Since $r \geq 2$, the Sobolev Theorem implies that the various maps defined on I are C^0. So the conditions specifying the various spaces are well defined.

In the light of Lemma 2.6 b), we take the inner product on $\mathcal{J}$ to be, for $\ell = r-2$,

$$(\xi,\xi)_{\ell,2} = (1/2)\left[(\mu,\mu)_{\ell,2}+\nu(1):\nu_2(1)+\nu(0):\nu_2(0)\right], \qquad (3.47)$$

where $(\mu,\mu)_{\ell,2}$ denotes the inner product on $W^{\ell,2}(I,L(E^3))$. The inner products on the other load spaces are defined similarly.

We extend the definition of the virtual work forms, the astatic load, and the convective representation for the loads in the two rod theories to this setting of generalized differentiability.

3.31 Definition. *Let k and r be as specified above and let $\mathcal{E}$ and $\mathcal{D}$ be the representation spaces given in Definition 2.21.*

a) *For $\xi \in \mathcal{J}$, $\gamma \in \mathcal{M}$, $\zeta \in \mathcal{F}$, and $Z \in \mathcal{E}$, define*

(1) *$\Omega(\xi,\gamma)$ to be the linear functional on $T_\gamma\mathcal{M}$ given by (3.31),*

(2) *$k(\xi,\gamma)$ to be the linear map on E^3 given by (3.32),*

(3) *$\beta(\xi,\gamma) \in \mathcal{F}$ by (3.35),*

(4) *$K(\zeta,Z)$ to be the bilinear form on $\mathcal{F} \times \mathcal{E}$ given by (3.36).*

b) *For $l \in \mathcal{L}$, $\chi \in \mathcal{N}$, $\Xi \in \mathcal{T}$, and $(b,Z) \in \mathcal{D}$, define*

(1) *$\Omega(l,\chi)$ to be the linear functional on $T_\chi\mathcal{N}$ given by (3.13),*

(2) *$k(l,\chi)$ to be the linear map on E^3 given by (3.14),*

(3) *$\beta(l,\chi) \in \mathcal{T}$ by (3.26),*

(4) *$K(\Xi,(b,Z))$ to be the bilinear form on $\mathcal{T} \times \mathcal{D}$ given by (3.27).* ∎

By the Sobolev Theorem the extensions are well defined. Moreover, Lemmas 3.9 and 3.17 extend.

3.32 Lemma. *Let k and r be as specified in Definition 3.30.*

a) *For the Kirchhoff theory,*

(1) *K is a nondegenerate bilinear form on $\mathcal{F} \times \mathcal{E}$;*

(2) *if $\xi \in \mathcal{J}$, $\gamma \in \mathcal{M}$, $H_\gamma \in T_\gamma\mathcal{M}$, $H_\gamma = \gamma Z$, $Z \in \mathcal{E}$, then*

$$\langle\Omega(\xi,\gamma),H_\gamma\rangle = K(\beta(\xi,\gamma),Z).$$

b) *For the special Cosserat theory,*

(1) *K is a nondegenerate bilinear form on $\mathcal{T} \times \mathcal{D}$;*

(2) *if $l \in \mathcal{L}$, $\chi \in \mathcal{N}$, $H_\chi \in T_\chi\mathcal{N}$, $H_\chi = L_\chi(b,Z)$, $(b,Z) \in D$,*

$$\langle\Omega(l,\chi),H_\chi\rangle = K(\beta(l,\chi),(b,Z)).$$

Proof. We sketch the proof of part b). In the proof of Lemma 3.9, the pair (b,Z) chosen to examine $\ker_1 K$ is C^∞; hence, it is an element of $\mathcal{D}$. Consequently, the proof of $\ker_1 K = 0$ extends to $\mathcal{T}$. Likewise, the proof of $\ker_2 K = 0$ extends to $\mathcal{N}$. Part a) follows by extending the proof of Lemma 3.16 in a similar manner. ∎

Finally, we extend the actions of Π_1 on the load spaces to the setting of extended differentiability.

3.33 Lemma.

a) The action of Π_1 on L or of Π or Π_s on J given by Definition 3.22 extend to actions on $\mathcal{L}$ and $\mathcal{J}$. Moreover, the actions are orthogonal with respect to the inner products on $\mathcal{L}$ and $\mathcal{J}$.

b) The representation of Π_1 on T or of Π or Π_s on F given by Definitions 3.25 and 3.28 extend to representations on $\mathcal{T}$ and $\mathcal{F}$. Moreover, the representations are orthogonal with respect to the inner products on $\mathcal{T}$ and $\mathcal{F}$.

Proof. By Lemma 2.18, if $l \in \mathcal{L}$ and $\xi \in \mathcal{J}$, gl and $g\xi$ given by (3.38) and (3.39) belong also to $\mathcal{L}$ and $\mathcal{J}$, respectively. Also by Lemma 2.18, if $\Xi \in \mathcal{T}$ and $\zeta \in \mathcal{F}$, then $\mathcal{A}(g)\Xi$ and $\mathcal{A}(g)\zeta$ do also. Carrying over the proof of Lemma 3.23 to the generalized differentiability setting then shows that the actions of Π_1 extend to $\mathcal{L}$, $\mathcal{J}$, $\mathcal{T}$, and $\mathcal{F}$. Since Π_1 acts orthogonally on E^3 and $L(E^3)$, a computation using (3.47) shows that

$$(g\xi, g\xi_2)_{\ell,2} = (\xi, \xi_2)_{\ell,2} \ . \tag{3.48}$$

Hence, Π_1 acts orthogonally on $\mathcal{J}$. Similar computations using the inner products of the other load spaces show that Π_1 acts orthogonally on them also. ∎

3.34 Lemma. *The representations of Π_1 on $\mathcal{T}$ and $\mathcal{D}$ or of Π and Π_s on $\mathcal{F}$ and $\mathcal{E}$ given by Lemma 3.33 satisfy (3.42), (3.43), (3.45), and (3.46).*

Proof. The proofs of Lemmas 3.26, 3.27, and 3.29 carry over to the generalized differentiability setting. ∎

IV. THE ROD EQUILIBRIUM PROBLEM

In this section we formulate the equilibrium problem for the special Cosserat and Kirchhoff rod theories as a variational problem on the space of loads and configurations specified previously. The model is a particular case of a Hamiltonian rod model in a convective representation of [33]. Assuming the material comprising the rod is hyperelastic, we introduce the potential function whose critical points determine the equilibrium configurations for the applied loads and boundary conditions. Assuming the material comprising the cross sections of the rod is isotropic and the cross sections themselves are circular, we indicate the symmetries which arise in the variational problem. Finally, for the Kirchhoff rod we extract from this general setting the particular problem of interest as a bifurcation problem.

Because of the functional analysis tools we will use to analyze the equilibrium problem we formulate the variational problem using the spaces of generalized differentiability introduced in Sections II.3 and III.4. For all that follows assume k is an integer, $k \geq 1$, r is an integer, and $r > k + 1/2$.

IV.1. *The Variational Functions*

The opportunity to formulate the equilibrium problem in a variational setting arises when we notice that we can derive the virtual work forms given in Definition 3.31 from a potential function.

4.1 Lemma.

a) For $(\xi,\gamma) \in \mathcal{J} \times \mathcal{M}$ *define* $trk(\xi,\gamma) = trace(k(\xi,\gamma)) \in \mathbb{R}$ *by*

$$trk(\xi,\gamma) = \int \mu : \gamma dS + \nu(1):\gamma(1) + \nu(0):\gamma(0), \tag{4.1}$$

where $k(\xi,\gamma)$ *is given in Definition 3.31. If* $H_\gamma \in T_\gamma \mathcal{M}$, *then*

$$\langle \Omega(\xi,\gamma), H_\gamma \rangle = [T_2(trk)(\xi,\gamma)]H_\gamma. \tag{4.2}$$

b) For $(l,\chi) \in \mathcal{L} \times \mathcal{N}$, *define* $trk(l,\chi) = trace(k(l,\chi)) \in \mathbb{R}$ *by*

$$trk(l,\chi) = \int_I [f\cdot x + \mathbb{Q}:\gamma]dS +$$
$$t(1)\cdot x(1) + \mathbb{P}(1):\gamma(1) + t(0)\cdot x(0) + \mathbb{P}(0):\gamma(0),$$

where $k(l,\chi)$ *is given by Definition 3.31. If* $(h,H_\gamma) \in T_\chi \mathcal{N}$,

$$\langle \Omega(l,x),(h,H_\gamma)\rangle = [T_2(trk)(l,x)](h,H_\gamma).$$

Proof. For $\xi \in \mathcal{J}$, trk$(\xi,\cdot)$ is the restriction to $\mathcal{M}$ of a linear functional on the Banach space $W^{r,2}(I,L(E^3))$. Consequently, $T_2(\text{trk})(\xi,\gamma) \equiv T(\text{trk}(\xi,\cdot))(\gamma)$ is the restriction of the linear functional to $T_\gamma \mathcal{M} \subseteq W^{r,2}(I,L(E^3))$. Part a) thus follows. Part b) follows similarly. ∎

4.2 Corollary.

a) Let $\xi \in \mathcal{J}$, $\gamma \in \mathcal{M}$, $H_\gamma \in T_\gamma \mathcal{M}$. *If* Z, $Z_2 \in W^{r,2}(I,skew(E^3))$ *are the convective and spatial representations for* H_γ *respectively, then*

$$[T_2(trk)(\xi,\gamma)]H_\gamma = K(\beta_2(\xi,\gamma),Z_2) = K(\beta(\xi,\gamma),Z). \tag{4.3}$$

b) Let $l \in \mathcal{L}$, $\chi = (x,\gamma) \in \mathcal{N}$ *and* $(h,H_\gamma) \in T_\chi \mathcal{N}$. *If* (b,Z) *and* (b_2,Z_2) *are the convective and spatial representations for* (h,H_γ), *then*

$$[T_2(trk)(l,\chi)](h,H_\gamma) = K(\beta_2(l,\chi),(b_2,Z_2)) = K(\beta(l,\chi),(b,Z)).$$

Here, β, β_2 *and* K *are given by Definition 3.31.*

Proof. The corollary follows by extending the proofs of Lemmas 4.1, 3.9, and 3.17 to the $W^{r,2}$ spaces. ∎

A variational formulation follows when we hypothesize that the material comprising the rod is hyperelastic and possesses an internal energy density function. Groups enter into the variational problem when we impose symmetries on the density function. Towards this end we adopt two hypotheses consistent with the approach of S. Antman [6].

4.3 Hypothesis.

a) *There is a smooth* (C^∞) *function* W *defined on* $I \times O(E^3) \times L(E^3)$ *which gives the total internal energy of the Kirchhoff rod in a configuration* $\gamma \in M$ *as*

$$E(\gamma) = \int_I W(S,\gamma(S),\gamma'(S))dS.$$

b) *There is a smooth* (C^∞) *function* W *defined on* $I \times E^3 \times GL(E^3) \times E^3 \times L(E^3)$ *which gives the total internal energy of the special Cosserat rod in a configuration* $\chi = (x,\gamma)$ *as*

$$E(\chi) = \int_I W(S,x(S),\gamma(S),x'(S),\gamma'(S))dS. \quad ■$$

4.4 Hypothesis. *For* $S \in I$ *assume the function* $W(S,\cdot)$ *satisfies the following conditions. Let* G_s *be given by Definition 2.4.*

a) *For the Kirchhoff rod: for* $A \in O(E^3)$, $B \in L(E^3)$,

$$(1) \quad W(S,QA,QB) = W(S,A,B) \quad \text{for all } Q \in O(E^3),$$

$$(2) \quad W(S,AQ^T,BQ^T) = W(S,A,B) \text{ for all } Q \in G_s.$$

b) *For the special Cosserat rod: for* $u,\ v \in E^3$, $A \in GL(E^3)$, $B \in L(E^3)$,

$$(1) \quad W(S,Qu + w,QA,Qv,QB) = W(S,u,A,v,B)$$

for all $w \in E^3$ *and* $Q \in O(E^3)$,

$$(2) \quad W(S,u,AQ^T,v,BQ^T) = W(S,u,A,v,B) \quad \text{for all } Q \in G_s. \quad ■$$

Remark. For the Kirchhoff rod, we may replace G_s by Γ in Hypothesis 4.4 without loss of generality. We seek a nontrivial restriction on a set of functions; namely, find those W for which

$$W(S,Q_1AQ_2^T,Q_1BQ_2^T) = W(S,A,B)$$

for all $(Q_1,Q_2) \in \Pi_1$. However, for $(Q_1,Q_2) = (-1,-1)$, the condition is satisfied by all functions W. Hence, Lemma 2.5 implies the restriction can be effectively imposed on all $(Q_1,Q_2) \in \Pi_1/H \approx \Pi = O(E^3) \times \Gamma$. We use this isomorphism, because we will be using the convective representation for the problem of interest.

In each part of the last hypothesis the first assumption is that of the Material Frame Indifference of the density. The second assumption is that of *transversal isotropy* ([9], p. 296f). For transversal isotropy, G_s (or Γ) represents the *rod material symmetry group,* which is a group reflecting the symmetry exhibited by the stress response of the material comprising the cross sections of the rod and the geometric shape of the cross sections. For example, if the material comprising the rod is assumed to be isotropic, but the cross sections of the rod are of triangular shape, the rod material symmetry group would be the dihedral group of order three, viewed as a subgroup of G_s ([4], p. 314), and the rod would not be transversely isotropic.

Hypothesis 4.4 allows us to express the internal energy density function in terms of functions invariant with respect to the group actions.

4.5 Lemma.

a) Let $u \in E^3$. Let $\tau(u) = (\tau_1(u),\tau_2(u))$, where

$$\tau_1(u) = \sum_{\alpha=1}^{2} u_\alpha u_\alpha, \text{ and } \tau_2(u) = u_3u_3,$$

for $u_j = u\cdot\underline{e}_j$, $j = 1, 2, 3$. Then there is a smooth (C^∞) function $H(S,\tau)$ for which

$$W(S,\gamma(S),\gamma'(S)) = H(S,\tau((u(\gamma))(S)),$$

for $\gamma \in M$, *let* $u(\gamma) \in W^{r-1,2}(I,E^3)$ *be given by*

$$\gamma'(S) = \gamma(S)(\hat{u}(\gamma))(S). \tag{4.4}$$

Moreover, for Π_1 *as given by Definition 2.4, for* $g \in \Pi_1$,

$$\tau((u(g\gamma))(S)) = \tau((u(\gamma))(S)).$$

b) *Let* $u,\ v \in E^3$, *let* $\lambda(u,v) = (\lambda_1(u,v),\lambda_2(u,v),\lambda_3(u,v))$, *where*

$$\lambda_1(u,v) = \tau_1(v),\ \lambda_2(u,v) = \tau_2(v),$$

and

$$\lambda_3(u,v) = \sum_{\alpha=1}^{2} u_\alpha v_\alpha.$$

Then there is a smooth (C^∞) *function* $H(S,\tau,\lambda)$ *for which*

$$W(S,x(s),\gamma(S),x'(S),\gamma'(S)) =$$
$$H(S,\tau((u(\chi))(S),\lambda((u(\chi))(S),(v(\chi))(S))),$$

for $\chi = (x,\gamma) \in N$, *and for* $u(\chi),\ v(\chi) \in W^{r-1,2}(I,E^3)$ *satisfying*

$$\gamma'(S) = \gamma(S)(\hat{u}(\chi))(S), \text{ and } x'(S) = \gamma(S)(v(\chi))(S).$$

Moreover, for Π *as given by Definition 2.4, for* $g \in \Pi$,

$$\tau((u(g\chi))(S)) = \tau((u(\chi)(S)),$$
$$\lambda((u(g\chi))(S),(v(g\chi))(S)) = \lambda((u(\chi))(S),(v(\chi))(S)).$$

Proof. The existence of the function H in each case is established in [9], p. 297 (see also [4], p. 315f). The invariance of τ and λ under the action of Π follows from their definition and Proposition 2.13. ∎

Remark. In general $\gamma' \notin T_\gamma \mathcal{M}$, because γ' is not a $W^{r,2}$ function. Rather, the "strains" (γ,γ') are elements of another space, the *first jet prolongation* of $\mathcal{M}$ ([17], p. 37). It is this space which is the space of strains for the Kirchhoff rod theory. As with the infinitesimal displacements, a strain can be given a convective and a spatial description in the *strain representation space* $W^{r-1,2}(I,O(E^3))$. The function $u(\gamma)$ of Lemma 4.5 is the convective representation of the strain, and is the strain variable which occurs in [33] (see [6]), for the case of the Kirchhoff rod. Similar remarks can be made for the description of strain for the case of the special Cosserat rod. To pursue this formulation for the strains, see [16].

From the hypotheses and Lemmas 4.1 and 4.5 we can construct the functionals for the variational problems.

4.6 Definition. *Given Hypotheses 4.3 and 4.4.*

a) If $\xi \in \mathcal{J}$ and $\gamma \in \mathcal{M}$, define the potential function for the Kirchhoff rod to be

$$\left.\begin{aligned} V(\xi,\gamma) &= E(\gamma) - trk(\xi,\gamma) \\ &= \int_I H(S,\tau((u(\gamma))(S))dS - trk(\xi,\gamma). \end{aligned}\right\} \quad (4.5)$$

b) If $l \in \mathcal{L}$, $\chi \in \mathcal{N}$, define the potential function for the special Cosserat rod to be

$$V(l,\chi) = E(\chi) - trk(l,\chi) =$$

$$\int_I H(S,\tau((u(\chi))(S),\lambda((u(\chi))(S),(v(\chi))(S)))dS - trk(l,\chi).$$

In either case, the function H is as specified in Lemma 4.5. ∎

The potential functions possess symmetries.

4.7 Lemma. *Let Π_1 be given by Definition 2.4, and let $g \in \Pi_1$. Then*

$$trk(g\xi, g\gamma) = trk(\xi, \gamma) \tag{4.6}$$
$$trk(gl, g\chi) = trk(l, \chi)$$

Proof. The lemma follows from (4.1) and the invariance of the inner product with respect to the action of Π_1 on $L(E^3)$. ∎

4.8 Proposition. *Let Π_1 be given by Definition 2.4, and let $g \in \Pi_1$.*

a) For the Kirchhoff rod,

$$V(g\xi, g\gamma) = V(\xi, \gamma). \tag{4.7}$$

b) For the special Cosserat rod,

$$V(gl, g\chi) = V(l, \chi).$$

Proof. Part a) follows from Lemmas 4.5 and 4.7. Part b) follows similarly. ∎

IV.2. *The Euler Field for the Kirchhoff Rod Problem*

The equilibrium configurations we seek for the Kirchhoff rod will be critical points of the potential function introduced in Definition 4.6. In this subsection we show that these critical points are zeros of a "vector field" on the manifold of configurations, the Euler field associated with the potential function. We derive the field and indicate the symmetries it possesses by virtue of Hypothesis 4.4.

4.9 Lemma. *Let $\gamma \in \mathcal{M}$, and let $u(\gamma)$ be given by (4.4). Then for $H_\gamma \in T_\gamma\mathcal{M}$ and for $Z \in \mathcal{C}$ satisfying $H_\gamma = T\mathcal{L}_\gamma(\gamma_0)Z$,*

$$[Tu(\gamma)]H_\gamma = Z' + [\hat{u}(\gamma), Z]. \tag{4.8}$$

If $Z(S) = \hat{z}(S)$, for $z \in W^{r,2}(I,E^3)$,

$$[Tu(\gamma)]H_\gamma = z' + u(\gamma) \times z.$$

Proof. Differentiating $\gamma' = \gamma\hat{u}(\gamma)$ with respect to γ gives

$$H'_\gamma = H_\gamma\hat{u}(\gamma) + \gamma[Tu(\gamma)]H_\gamma.$$

Now $H_\gamma = \gamma\cdot Z$ implies

$$H'_\gamma = \gamma' Z + \gamma Z' = \gamma\hat{u}(\gamma)Z + \gamma Z'.$$

So

$$\gamma[Tu(\gamma)]H_\gamma = \gamma Z' + \gamma\hat{u}(\gamma)Z - \gamma Z\hat{u}(\gamma) = \gamma\{Z' + [\hat{u}(\gamma),Z]\},$$

giving the first result. Lemma 2.6 and the above computation give the second result. ∎

The Euler field arises when we express the derivative of the potential function in terms of the binear form K. Towards this end we define the following operator.

4.10 Definition. *Given Hypotheses 4.3 and 4.4,*

a) for $S \in I$ and $u \in E^3$, define $m(S,u) \in E^3$ by

$$m(S,u) = 2\,\frac{\partial H}{\partial \tau_1}(S,\tau(u))u_\alpha \underline{e}_\alpha + 2\,\frac{\partial H}{\partial \tau_2}(S,\tau(u))u_3\underline{e}_3. \tag{4.9}$$

b) For $\gamma \in \mathcal{M}$ set $\Phi(\gamma) \equiv (\hat{q}, \hat{p}) \in \mathcal{F}$, for

$$\left.\begin{aligned} q(S) &= -[m(S,(u(\gamma))(S))]' - (u(\gamma))(s) \times m(S,(u(\gamma))(S)),\\ p(1) &= m(1,(u(\gamma))(1)), \quad p(0) = -m(0,(u(\gamma))(0)), \end{aligned}\right\} \tag{4.10}$$

where $u(\gamma)$ is given by (4.4). ∎

From Lemma 4.5 it follows that Φ is a smooth (C^∞) function of γ.

4.11 Proposition. *Let $\gamma \in \mathcal{M}$ and $H_\gamma \in T_\gamma\mathcal{M}$. Let $Z \in \mathcal{C}$ be the convective representation for H_γ. Let $E(\gamma)$ be given as in Hypothesis 4.3. Then*

$$[TE(\gamma)]H_\gamma = K(\Phi(\gamma),Z),$$

where K and Φ are given by Definitions 3.31 and 4.10.

Proof. Write $H(S,\tau((u(\gamma))(S)))$ as the composition of maps

$$H(S,\tau((u(\gamma))(S))) = \{(\mathcal{H}\circ\tau\circ u)(\gamma)\}(S).$$

Then

$$[TE(\gamma)]H_\gamma = \int_I \{D\mathcal{H}(\tau(u(\gamma)))][T\tau(u(\gamma))][Tu(\gamma)]H_\gamma\}(S)dS.$$

For $w \in W^{r-1,2}(I,E^3)$ the definition of τ and Lemma 2.18 imply

$$[T\tau(u(\gamma))]w = (\ 2(u(\gamma))_\alpha w_\alpha,\ 2(u(\gamma))_3 w_3\ ) \in W^{r-1,2}(I,R^2).$$

Lemmas 2.6 and 4.5 and (4.9) give

$$\{[T\mathcal{H}(\tau(u(\gamma)))][T\tau(u(\gamma))]w\}(S) =$$

$$2\left(\frac{\partial H}{\partial \tau_1}(S,\tau((u(\gamma))(S)))\right)((u(\gamma))(s))_\alpha\, w_\alpha +$$

$$2\left(\frac{\partial H}{\partial \tau_2}(S,\tau((u(\gamma))(S)))\right)((u(\gamma))(S))_3\, w_3 =$$

$$m(S,(u(\gamma))(S))\cdot w(S) = (1/2)\hat{m}(S,(u(\gamma))(S)):\hat{w}(S).$$

Taking $w = [Tu(\gamma)]H_\gamma$, (4.8) gives

$$\{[T\mathcal{H}(\tau(u(\gamma)))][T\tau(u(\gamma))][Tu(\gamma)]H_\gamma\}(S) =$$

$$(1/2)\hat{m}(S,(u(\gamma))(S)):(Z'(S) + [(\hat{u}(\gamma)(S),Z(S)]).$$

Lemma 2.6 implies

$$\hat{m}(S,(u(\gamma))(S)):[\hat{u}(\gamma)(S),Z(S)] = -[(\hat{u}(\gamma))(S),\hat{m}(S,(u(\gamma))(S))]:Z(S)$$
$$= -((u(\gamma))(S) \times m(S,(u(\gamma))(S))):Z(S).$$

So $[TE(\gamma)]H_\gamma =$

$$(1/2)\int_I(\hat{m}(S,(u(\gamma))(S)):Z'(S)-((u(\gamma))(S) \times m(S,(u(\gamma))(S))):Z(S))dS.$$

Integrating the first term by parts and using the definitions of K and Φ give the result. ∎

4.12 Theorem. *Let $\xi \in \mathcal{J}$, $\gamma \in \mathcal{M}$, and $H_\gamma \in T_\gamma\mathcal{M}$. Let Z be the convective representation for H_γ. Let V be given by (4.5). Then*

$$[T_2V(\xi,\gamma)]H_\gamma = K(\Psi(\xi,\gamma),Z), \tag{4.11}$$

where Ψ is the smooth (C^∞) map given by

$$\Psi(\xi,\gamma) = \Phi(\gamma) - \beta(\xi,\gamma) \in \mathcal{F}, \tag{4.12}$$

and where K, β, and Φ are given by Definitions 3.31 and 4.10.

Proof. The result follows from Proposition 4.11, (4.3), and the remark following Definition 4.10. ∎

Call $\Psi(\xi,\gamma)$ the *convective representation of the Euler field* associated with the potential V. Call Φ the *convective representation of the elastostatic operator.* The relation of these operators to the differential equations of equilibrium will be developed in the next subsection.

The symmetries which $V(\xi,\gamma)$ possesses impose symmetries on $\Psi(\xi,\gamma)$ and $\Phi(\gamma)$.

4.13 Proposition. *Let $\xi \in \mathcal{J}$, and $\gamma \in \mathcal{M}$. Let Φ and Ψ be as given in (4.10) and (4.12). Let Π be given by Definition 2.4. Then for $g \in \Pi$,*

$$\Phi(g\gamma) = \mathcal{A}(g)\Phi(\gamma), \tag{4.13}$$

$$\Psi(g\xi, g\gamma) = \mathcal{A}(g)\Psi(\xi, \gamma), \tag{4.14}$$

where $\mathcal{A}(g)$ is given by Lemma 3.33.

Proof. From Lemma 4.5, $E(g\gamma) = E(\gamma)$. So for $H_\gamma \in T_\gamma \mathcal{M}$, the chain rule and Proposition 2.13 give

$$[TE(\gamma)]H_\gamma = [TE(g\gamma)]g \cdot H_\gamma.$$

Proposition 4.11, (2.23), and (3.45) then imply

$$K((\Phi(g\gamma), \mathcal{B}(g)Z) = K(\Phi(\gamma), Z) = K(\mathcal{A}(g)\Phi(\gamma), \mathcal{B}(g)Z).$$

As Z is arbitrary, the nondegeneracy of K gives (4.13). By (4.7), the chain rule, (4.11), and (2.23),

$$K(\Psi(g\xi, g\gamma), \mathcal{B}(g)Z) = K(\Psi(\xi, \gamma), Z).$$

Equation (3.45) and the nondegeneracy of K then give (4.14). ∎

IV.3. *The Constrained Equilibrium Problem*

The equilibrium problem for the Kirchhoff rod which we wish to consider arises when we constrain the left end of the centerline to remain fixed at the origin and require the right end to lie along the $\underline{e}_3$ axis. We formulate it in terms of the elements introduced in Sections II.3 and IV.2 this way.

4.14 Problem. *Let $\mathcal{M}_1$ be as specified in Definition 2.23. Let V be given by (4.5). Let V_1 be the restriction of V to $\mathcal{J} \times \mathcal{M}_1$. For $\xi \in \mathcal{J}$ find the set of critical points for $V_1(\xi, \cdot)$.* ∎

By Proposition 2.24, $\mathcal{M}_1$ is a closed submanifold of $\mathcal{M}$. So the problem is one of finding extrema of $V(\xi, \cdot)$ subject to a constraint. Consequently, we may formulate the problem using Lagrange multipliers.

4.15 Problem. *For $\alpha = 1, 2$, for $\gamma \in \mathcal{M}$, define $\iota_\alpha(\gamma) \in \mathbb{R}$ by*

$$\iota_\alpha(\gamma) = \int_I \gamma(S) : \underline{e}_\alpha \otimes \underline{e}_3 \, dS.$$

For $c_1, c_2 \in R$ define $V(\xi, \gamma, c_1, c_2) \in R$ by

$$V(\xi, \gamma, c_1, c_2) = V(\xi, \gamma) - c_\alpha \iota_\alpha(\gamma).$$

For $\xi \in \mathcal{J}$ find the set of critical points for $V(\xi, \cdot, \cdot, \cdot)$. ∎

In this subsection we will derive the differential equations and boundary conditions for equilibrium, and show the equivalence of the two problems. To do so, we decompose the representation spaces for the infinitesimal displacements and the applied loads in a manner which is consistent with the constraint. We will also use this decomposition in Section V to reduce the problem as stated to one involving a finite number of degrees of freedom.

To obtain the decomposition we desire we first identify the convective representation for those infinitesimal displacements at $\gamma \in \mathcal{M}_1$ which maintain the constraint defining $\mathcal{M}_1$. We then use it to decompose the representation space for the unconstrained infinitesimal displacements.

4.16 Lemma. *For $\alpha = 1, 2$, let $\xi_\alpha \in \mathcal{J}$ be given by $\xi_\alpha \equiv (\mu, \nu)$,*

$$\mu(S) \equiv \underline{e}_\alpha \otimes \underline{e}_3 \, , \quad \nu(S) \equiv 0 \, .$$

Let k, β, and K be given by Definition 3.31. Let $\mathrm{i} : \mathcal{M} \longrightarrow \mathbb{R}^2$ be defined by

$$\mathrm{i}(\gamma) = trk(\xi_\alpha, \gamma)\underline{e}_\alpha \, .$$

a) *If $\mathcal{C}$ is given by Definition 2.21, if $Z \in \mathcal{C}$, then*

$$T\mathrm{i}(\gamma)T\mathcal{L}_\gamma(\gamma_0)Z = K(\beta(\xi_\alpha, \gamma), Z)\underline{e}_\alpha \, . \tag{4.15}$$

b) $\gamma \in M_1$ if and only if $i(\gamma) = 0$.

Proof. The lemma follows from (4.3) and Proposition 2.24. ∎

4.17 Lemma. *Let $\mathfrak{C}$ be the representation space given in Definition 2.21. Then $\mathfrak{C}$ is an inner product space relative to the bilinear form*

$$B(Z_1, Z_2) = (1/2) \int_I Z_1(S) : Z_2(S) dS. \qquad (4.16)$$

Proof. Since U:V is an inner product on skew(E^3), it follows that $B(Z,Z) = 0$ if and only if $Z \equiv 0$. Hence, B is a nondegenerate bilinear form. ∎

The bilinear form B is the *Betti form* for the representation space $\mathfrak{C}$ (see [2], p. 368).

4.18 Proposition. *For $\alpha = 1$, 2, let $\xi_\alpha \in \mathcal{J}$ be given as in Lemma 4.16. Let $\gamma \in M_1$. Define $\eta_\alpha(\gamma) \in \mathfrak{C}$ by*

$$\big(\eta_\alpha(\gamma)\big)(S) = skew(\gamma^T(S)\underline{e}_\alpha \otimes \underline{e}_3).$$

Define the subspaces $S_{1\gamma}$, $S_{2\gamma}$ of $\mathfrak{C}$ by

$$\left. \begin{array}{l} S_{1\gamma} = span_\mathbb{R}\{\ \eta_\alpha(\gamma),\ \alpha = 1,\ 2\ \}, \\[2mm] S_{2\gamma} = \{\ Z \in \mathfrak{C}\ |\ K(\beta(\xi_\alpha, \gamma), Z) = 0,\ \alpha = 1,\ 2\ \}. \end{array} \right\} \quad (4.17)$$

Then a) $dimS_{1\gamma} = 2$.
  b) $\mathfrak{C} = S_{1\gamma} \oplus S_{2\gamma}$,
  a direct sum which is orthogonal relative to B of (4.16).
  c) For $\mathcal{L}_\gamma$ given by Corollary 2.19, for $\gamma_0 \in M_1$ given by $\gamma_0(S) \equiv 1$,

$$\left.\begin{aligned}&[T\mathcal{L}_\gamma(\gamma_0)]S_{2\gamma} = T_\gamma \mathcal{M}_1,\\&T_\gamma\mathcal{M} = [T\mathcal{L}_\gamma(\gamma_0)]S_{1\gamma} \oplus [T\mathcal{L}_\gamma(\gamma_0)]S_{2\gamma}\ .\end{aligned}\right\} \quad (4.18)$$

Proof.

a) We show that the $\eta_\alpha(\gamma)$ are linearly independent for each $\gamma \in \mathcal{M}_1$. If for some $\gamma \in \mathcal{M}_1$ and $\lambda \in R$, $\eta_1(\gamma) = \lambda\eta_2(\gamma)$, then $(1 + \lambda^2)\gamma_{13}(S) \equiv 1$, for $\gamma_{13}(S) = \underline{e}_1\cdot\gamma(S)\underline{e}_3$. So, $i(\gamma) \neq 0$, a contradiction by Lemma 4.16.

b) By the definitions of the $\eta_\alpha(\gamma)$, the ξ_α, K, and B, if $Z \in \mathcal{E}$,

$$K(\beta(\xi_\alpha,\gamma),Z) = B(\eta_\alpha,Z).$$

Hence, Lemma 4.18 implies that $S_{2\gamma}$ is a direct summand for $S_{1\gamma}$ in $\mathcal{E}$ which, by construction, is orthogonal relative to B.

c) Since $\mathcal{L}_\gamma$ is a diffeomorphism of $\mathcal{M}$, $T\mathcal{L}_\gamma(\gamma_0)$ is an isomorphism between the tangent spaces $T_{\gamma_0}\mathcal{M} = \mathcal{E}$ and $T_\gamma\mathcal{M}$. Since $\mathcal{M}_1 = i^{-1}(0)$, for $Z \in \mathcal{E}$, $T\mathcal{L}_\gamma(\gamma_0)Z \in T_\gamma\mathcal{M}_1$ if and only if

$$Ti(\gamma)T\mathcal{L}_\gamma(\gamma_0)Z = 0\ , \qquad (4.19)$$

or by (4.15), $K(\beta(\xi_\alpha,\gamma),Z) = 0$, $\alpha = 1, 2$. Hence,

$$T\mathcal{L}_\gamma(\gamma_0)S_{2\gamma} = T_\gamma\mathcal{M}_1\ .$$

Extend B to a (left invariant) metric on $\mathcal{M}$ by defining

$$B_\gamma(H_{1\gamma},H_{2\gamma}) = B(Z_1,Z_2),$$

for $\gamma \in \mathcal{M}$, $H_{1\gamma}$, $H_{2\gamma} \in T_\gamma\mathcal{M}$, and $H_{\alpha\gamma} = \gamma\cdot Z$, $\alpha = 1, 2$. Relative to this metric $[T\mathcal{L}_\gamma(\gamma_0)]S_{1\gamma}$ and $[T\mathcal{L}_\gamma(\gamma_0)]S_{2\gamma}$ remain orthogonal. Thus,

$$T_\gamma\mathcal{M} = [T\mathcal{L}_\gamma(\gamma_0)](S_{1\gamma} \oplus S_{2\gamma}) = [T\mathcal{L}_\gamma(\gamma_0)]S_{1\gamma} \oplus T_\gamma\mathcal{M}_1,$$

and the sum is orthogonal relative to B. ∎

The space $S_{2\gamma}$ is the space of convective representations for the infinitesimal displacements at $\gamma \in \mathcal{M}_1$ which maintain the constraint. It is important to notice that this subspace of $\mathcal{E}$ *changes* as γ varies in $\mathcal{M}_1$. The sum (4.18) is a decomposition of the space of unconstrained infinitesimal displacements at γ, $T_\gamma\mathcal{M}$, in a manner consistent with the constraint.

Remark. The variation of $S_{2\gamma}$ with γ manifests the fact that $\mathcal{M}_1$ is *not* a Lie subgroup of $\mathcal{M}$. Indeed, were this the case, $S_{2\gamma}$ would be a constant subspace of $\mathcal{E}$, and in fact a representation space for the constrained infinitesimal displacements. In consequence, the reduction of the problem of interest to one with a finite number of degrees of freedom is not a trivial exercise.

We may decompose the representation space for the applied loads in an analogous manner.

4.19 Lemma. *Let $\gamma \in \mathcal{M}_1$ Define the subspaces $\mathcal{F}_{1\gamma}$ and $\mathcal{F}_{2\gamma}$ of $\mathcal{F}$ by*

$$\left.\begin{aligned} \mathcal{F}_{1\gamma} &= span_{\mathbb{R}}\{\zeta_\alpha(\gamma) \equiv \beta(\xi_\alpha,\gamma),\ \alpha = 1,\ 2\} \\ \mathcal{F}_{2\gamma} &= \{\zeta \in \mathcal{F} \mid K(\zeta,Z) = 0 \text{ for } Z \in S_{1\gamma}\}. \end{aligned}\right\} \quad (4.20)$$

Then a) $\mathcal{F} = \mathcal{F}_{1\gamma} \oplus \mathcal{F}_{2\gamma}$,

b) $K(\zeta,Z) = 0$ for $Z \in S_{2\gamma}$ if and only if $\zeta \in \mathcal{F}_{1\gamma}$.

Proof. The lemma follows from the nondegeneracy of K on $\mathcal{E} \times \mathcal{F}$, the definitions of the subspaces involved, and the relation between K and B established in the proof of Proposition 4.18. ∎

The equations for the constrained equilibria and the equivalence of the two problems now follow.

4.20 Theorem. *Let $\xi \in \mathcal{J}$.*

a) γ is a critical point for $V_1(\xi,\cdot)$ of Problem 4.14 if and only if

$$\Psi(\xi,\gamma) \in \mathcal{F}_{1\gamma}.$$

b) γ *is a critical point for* $V_1(\xi,\cdot)$ *if and only if for some* $c_\alpha \in R$, $\alpha = 1, 2$, γ *solves the boundary value problem*

$$\hat{m}' + u \times m + \hat{q} = c_\alpha \eta_\alpha(\gamma), \quad \text{for } S \in (0,1),$$

$$\hat{m}(1) - \hat{p}(1) = 0, \text{ and } \hat{m}(0) + \hat{p}(0) = 0,$$

where $(\hat{q},\hat{p}) = \beta(\xi,\gamma)$, u *and* m *are as given in (4.4) and (4.9), respectively, and* $\eta_\alpha(\gamma)$ *is given by Proposition 4.18.*

c) Problems 4.14 and 4.15 are equivalent.

Proof.

a) By Theorem 4.12 and Proposition 4.18, if $\gamma \in \mathcal{M}_1$, $H_\gamma \in T_\gamma\mathcal{M}_1$,

$$[TV_1(\xi,\cdot)(\gamma)]H_\gamma = K(\Psi(\xi,\gamma),Z),$$

where $Z \in S_{2\gamma}$ is the convective representation for H_γ. So Lemma 4.19 implies that γ is a critical point for $V_1(\xi,\cdot)$ if and only if $\Psi(\xi,\gamma) \in \mathcal{F}_{1\gamma}$.

b) This part follows from the definition of $\Psi(\xi,\gamma)$ and $\mathcal{F}_{1\gamma}$.

c) For $(\gamma,c_1,c_2) \in \mathcal{M} \times \mathbb{R}^2$ and $(Z,r_1,r_2) \in \mathcal{E} \times \mathbb{R}^2$, Problem 4.15 gives

$$\left.\begin{aligned} &[T\mathcal{V}(\xi,\cdot,\cdot,\cdot)(\gamma,c_1,c_2)](H_\gamma,r_1,r_2) \\ &= [TV(\xi,\cdot)H_\gamma - r_\alpha i_\alpha(\gamma) - c_\alpha K((\beta(\xi_\alpha,\gamma),Z) \\ &= K(\Psi(\xi,\gamma),Z) - r_\alpha i_\alpha(\gamma) - c_\alpha K((\beta(\xi_\alpha,\gamma),Z) \\ &= K(\Psi(\xi,\gamma) - c_\alpha \zeta_\alpha(\gamma),Z) - r_\alpha i_\alpha(\gamma), \end{aligned}\right\} \quad (4.21)$$

If $\gamma \in \mathcal{M}_1$, $i_\alpha(\gamma) = 0$. If γ is a critical point for $V_1(\xi,\cdot)$, choosing the c_α as indicated by part b) implies that the right hand side of (4.21) is zero, and that (γ,c_1,c_2) is a critical point for $\mathcal{V}(\xi,\cdot,\cdot,\cdot)$. Conversely, if (γ,c_1,c_2) is a critical point for $\mathcal{V}(\xi,\cdot,\cdot,\cdot)$, the right hand side of (4.18) is zero for arbitrary (Z,r_1,r_2). As K is nondegenerate,

$$i_\alpha(\gamma) = 0, \text{ and } \Psi(\xi,\gamma) \in \mathcal{F}_{1\gamma}.$$

So $\gamma \in \mathcal{M}_1$, and γ is a critical point for $V_1(\xi,\cdot)$ by part b). ∎

Part b) of Theorem 4.20 gives the differential equations and boundary conditions for the constrained equilibria. Our interest will be in the formulation of the problem as presented in part a).

We conclude this subsection by indicating how we can relate the Lagrange multipliers c_α arising in Theorem 4.20 to the forces maintaining the constraints of Problem 4.14. For $\alpha = 1, 2$, we can use (2.23) and (3.35) to identify a load in $\mathcal{J}$ which produces $c_\alpha\zeta_\alpha(\gamma)$ of (4.21) in the configuration γ. Taking the constraint forces to act at the ends of the rod, set

$$f(S) \equiv 0, \quad \mathbb{Q}(S) \equiv 0,$$

$$t(1) = c_\alpha \underline{e}_\alpha = -t(0).$$

$$\mathbb{P}(1) = \mathbb{P}(0) = 0.$$

By (3.23), $\xi = (\mu,\nu)$,

$$\mu(S) = t(1) \otimes \underline{e}_3 \equiv c_\alpha \underline{e}_\alpha \otimes \underline{e}_3,$$

$$\nu(S) = 0.$$

Thus $\beta(\xi,\gamma) = c_\alpha\zeta_\alpha(\gamma)$ and we may view $c_\alpha\zeta_\alpha(\gamma)$ as arising from tractions which are present at the ends of the rod, and which lie in the $(\underline{e}_1,\underline{e}_2)$ plane. These are the tractions generated by the constraints which keep the ends of the rod on the $\underline{e}_3$ axis. The component of the traction in each of the two directions is the associated Lagrange multiplier.

IV.4. *The Bifurcation Problem for the Kirchhoff Rod*

We now formulate the problem of interest as a bifurcation problem arising from the equilibrium problem for the Kirchhoff rod formulated in the previous subsection. Under the previous assumptions the straight configuration is an equilibrium configuration for the rod under an axially symmetric compressive load. This configuration generates an orbit of equilibrating configurations for the load. We formulate as a bifurcation problem the question of how this orbit is altered when we perturb the compressive load by additional loads which break the axial symmetry.

Let $p > 0$. As indicated in Example 3.18, $\xi_0 = (\mu,0) \in \mathcal{J}$ for $\mu(S) = -p\, \underline{e}_3 \otimes \underline{e}_3$ characterizes an axially symmetric compressive load for the Kirchhoff rod. In the following definition we formally recognize it and the subgroup of Π which leaves both it and the manifold $\mathcal{M}_1$ of constrained configurations invariant.

4.21 Definition. *Let $\xi_3 = (\mu,0) \in \mathcal{J}$, for $\mu(S) = -\underline{e}_3 \otimes \underline{e}_3$.*

a) Call

$$\{\; p\xi_3 \mid p \in \mathbb{R} \;\} \subseteq \mathcal{J}$$

the (1-parameter) family of unperturbed loads *for the problem of interest.*

b) Let $\mathcal{G}$ be given by (2.42). Set

$$G = \{\; g \in \mathcal{G} \mid g\xi_3 = \xi_3 \;\},$$

where the action of $\mathcal{G}$ on $\mathcal{J}$ is given by Lemma 3.33. Call G the load isotropy group *for ξ_3 (hence, $p\xi_3$, $p \in \mathbb{R}$).* ■

Notation. When there is no confusion we will write $\xi_0 = p\xi_3$. ■

4.22 Lemma. *Let J and Σ_3 be given by Definition 2.4. Let G be the group given by Definition 4.21. Then*

$$G = \big(O(2) \times SO(2)\big) <(\Sigma_3, J\Sigma_3)>,$$

a semidirect product group in $\mathcal{G}$.

Proof.

Definition 2.4 and a computation establishes that

$$\Gamma = SO(2)\langle J\Sigma_3\rangle, \qquad (4.22)$$

a semidirect product subgroup of G_s. Since $Q \in O(2)$ implies $Q\underline{e}_3 = \underline{e}_3$, and since $\Sigma_3\underline{e}_3 = -\underline{e}_3$, G contains the subgroups $(O(2) \times SO(2))$ and $\langle(\Sigma_3, J\Sigma_3)\rangle$ of $\mathcal{G}$. A computation shows that $(O(2) \times SO(2))$ is a normal subgroup of G. Hence, $(O(2) \times SO(2))\langle(\Sigma_3, J\Sigma_3)\rangle$, a semidirect product, is a subgroup of G. Conversely, the definition implies that G is generated by $(O(2) \times SO(2))$ and $(\Sigma_3, J\Sigma_3)$. Hence, G is contained in the smallest subgroup generated by them, which is $(O(2) \times SO(2))\langle(\Sigma_3, J\Sigma_3)\rangle$. ∎

We now recognize that the initially straight configuration for the rod is a stress-free configuration, and that it is an equilibrating configuration for an axially symmetric compressive load.

4.23 Lemma. *Let $\gamma_0 \in \mathcal{M}_1$ be specified by $\gamma_0(S) \equiv 1 \in O(E^3)$. Then*

a) $\Phi(\gamma_0) = 0 \in \mathcal{F}$;

b) γ_0 is a critical point for $V_1(\xi_0,\cdot)$ of Problem 4.14.

Proof. Since $u(\gamma_0) \equiv 0$, part a) follows from the definition of Φ. By (3.35) $\beta(\xi_0,\gamma_0) = 0 \in \mathcal{F}$. Part b) then follows from Theorem 4.20. ∎

Terminology. Call γ_0 the *trivially equilibrating configuration* for the unperturbed load ξ_0.

By the symmetry which the variational problem exhibits γ_0 generates an orbit of equilibrating configurations for the unperturbed load.

4.24 Lemma. *Let $V_1(\xi_0,\cdot)$ be as specified in Problem 4.14. Let $\mathrm{crit}V_1(\xi_0,\cdot)$ denote the set of critical points for $V_1(\xi_0,\cdot)$ in $\mathcal{M}_1$. Let*

$$G\gamma_0 = \{\ \gamma \in \mathcal{M}_1 \mid \gamma = g\gamma_0,\ g \in G\ \}. \qquad (4.23)$$

Then
$$G\gamma_0 \subseteq \mathrm{crit}\ V_1(\xi_0,\cdot).$$

Proof. If $g \in G$, (4.7) gives

$$V_1(\xi_0,\gamma_0) = V_1(g\xi_0,g\gamma_0) = V_1(\xi_0,g\gamma_0).$$

Consequently for $Z \in T_{\gamma_0}\mathcal{M}_1$, the chain rule and Lemma 2.22 give

$$[T(V_1(\xi_0,\cdot)(\gamma_0)]Z = [T(V_1(\xi_0,\cdot)(g\gamma_0)]g\cdot Z.$$

Since left translation by $g \in G$ induces an isomorphism between tangent spaces, $\gamma_0 \in \mathrm{crit}V_1(\xi_0,\cdot)$ if and only if $g\gamma_0 \in \mathrm{crit}V_1(\xi_0,\cdot)$. ∎

Terminology. Call $G\gamma_0$ the *orbit of trivially equilibrating configurations* for the unperturbed load ξ_0.

The question we pose is to determine if the orbit $G\gamma_0$ bifurcates when the unperturbed load ξ_0 is perturbed in a prescribed way. That is to say, if we perturb the load ξ_0 to a new load ξ we seek to determine an orbit of equilibrating configurations for it which approaches the trivial orbit $G\gamma_0$ in $\mathcal{M}_1$ as ξ approaches ξ_0 in $\mathcal{J}$. If the orbit for ξ is *not* in one-to-one correspondence with $G\gamma_0$, we say the trivial orbit *bifurcates* under the perturbation. If the orbit for ξ *is* in one-to-one correspondence with $G\gamma_0$, we say *no bifurcation occurs* under the perturbation.

We formulate the question mathematically this way.

4.25 Problem. *Let $U(G\gamma_0)$ be a given neighborhood of $G\gamma_0$ in $\mathcal{M}_1$. For ξ in a neighborhood of ξ_0 in $\mathcal{J}$, determine the sets*

$$\mathrm{crit}V_1(\xi,\cdot) \cap U(G\gamma_0) \quad \text{and} \quad \mathrm{crit}V_1(\xi_0,\cdot) \cap U(G\gamma_0). \quad ∎$$

Problem 4.25 is a bifurcation problem in the sense that we are seeking to determine how a set of critical points of a function changes as a parameter changes. To aid our analysis, we identify a subgroup of G which will be important, and decompose G in terms of it.

4.26 Definition.

a) Set

$$\hat{\Gamma} = \{\ g \in G \mid g\gamma_0 = \gamma_0\ \}. \tag{4.24}$$

Call $\hat{\Gamma}$ *the* isotropy group *for the trivially equilibrating configuration* γ_0.

b) If $g \in G$, *set*

$$g\hat{\Gamma} = \{\ gh^{-1} \mid h \in \hat{\Gamma}\ \}. \tag{4.25}$$

Call $g\hat{\Gamma}$ *the* $\hat{\Gamma}$ orbit in G through g, *or the* right coset of g by $\hat{\Gamma}$. ∎

4.27 Proposition. *Let* Γ *be given by Definition 2.4. Set*

$$G_1 = \{\ g \mid g = (Q,1),\ Q \in O(2)\ \}, \tag{4.26}$$

Then

a) $\hat{\Gamma} = \Gamma \Delta \Gamma \equiv \{\ (Q,Q) \in G \mid Q \in \Gamma\ \}.$ (4.27)

b) $G = G_1\hat{\Gamma},$ (4.28)

a semidirect product.

Proof.

a) A computation shows that $\Gamma \Delta \Gamma$ is a subgroup of $\hat{\Gamma}$. Conversely, let $g = (Q_1,Q_2) \in \hat{\Gamma}$. Since $g \in G$, $Q_1 = R_1js$, and $Q_2 = R_2t$, where $R_1, R_2 \in SO(2)$, $j \in \langle J\rangle$, $s \in \langle\Sigma_3\rangle$, and $t \in \langle J\Sigma_3\rangle$. Since $g\gamma_0 = \gamma_0$, $1 = R_1jstR_2^{-1}$. Since $1 \in SO(E^3)$ and $1\underline{e}_3 = \underline{e}_3$, $js = t$. Hence, $R_1 = R_2$, and $Q_1 = Q_2 \in SO(2)\langle J\Sigma_3\rangle = \Gamma$.

b) Since $G = \big(O(2) \times SO(2)\big)\langle(\Sigma_3, J\Sigma_3)\rangle$, a computation shows that G_1 is a normal subgroup of G. Hence $G_1\hat{\Gamma}$, the semidirect product, is a subgroup of G. Conversely, an argument analogous to that of part a) establishes that G is a subgroup of $G_1\hat{\Gamma}$. ∎

Remark. If $R_\phi \in SO(2)$ represents a rotation about the $\underline{e}_3$ axis through an angle ϕ, a computation shows that $R_\phi JR_{-\phi} = R_{2\phi}$. Hence, $\langle J\rangle$ is not a normal subgroup of O(2), and consequently, $\hat{\Gamma}$ is not a normal subgroup of G. So $G_1\hat{\Gamma}$ is a semidirect product subgroup of

G, but not an internal direct product.

In the development of Section VI, we will use the subgroups of G and G_1 which contain the identity, and the commutation relations between the subgroups G_1 and $\hat{\Gamma}$. We present these features in the next two lemmas.

4.28 Lemma. *Let SG (SG_1) denote the subgroup of G (G_1) which contains the identity element.*

a) Then

$$SG_1 = \{ g \mid g = (Q,1),\ Q \in SO(2) \},$$

and
$$SG = SG_1 S\hat{\Gamma},$$

a semidirect product, where

$$S\hat{\Gamma} = SO(2) \Delta SO(2) .$$

b) As manifolds,

i) $\hat{\Gamma}$ is a disjoint union of two circles,
ii) SG is a torus,
iii) $SG_1\hat{\Gamma}$ is a disjoint union of two tori,
iv) G_1 is a disjoint union of two circles,
v) G is a disjoint union of four tori.

Proof.
a) SG_1 follows from (4.26). From (4.27), $S\hat{\Gamma}$ is the component of $\hat{\Gamma}$ containing the identity. The representation of SG as a semidirect product then follows.
b) As a manifold SO(2) is diffeomorphic to a circle. As a union of cosets $O(2) = SO(2) \cup SO(2)J$. Since $\hat{\Gamma}$ is isomorphic to Γ, which in turn is isomorphic to O(2), the descriptions of $\hat{\Gamma}$ and G_1 follow. As topological spaces the semidirect products are Cartesian products. Hence, the descriptions of SG and $SG_1\hat{\Gamma}$ follow. Finally, as a union of cosets $G = SG_1\hat{\Gamma} \cup (J,1)SG_1\hat{\Gamma}$. Hence, the description of G follows. ∎

4.29 Lemma. *If $\sigma \in G_1$ and $g \in \hat{\Gamma}$, set*

$$Ad_g\sigma = g\sigma g^{-1}, \tag{4.29}$$

where the product is taken in G. Then $Ad_g\sigma \in G_1$.

a) If $\sigma \in SG_1$,

$$Ad_g\sigma = \begin{cases} \sigma & \text{if } g \in S\Gamma \\ \sigma^{-1} & \text{if } g \in S\Gamma(J\Sigma_3, J\Sigma_3) \end{cases} .$$

b) If $\sigma = (R_\phi, J, 1) \in SG_1(J,1)$, *if* $g = (Q,Q) \in \Gamma$,

$$Ad_g\sigma = \begin{cases} (R_{2\psi}, 1)\sigma & \text{if } Q = R_\psi \\ (R_{-2\psi}, 1)\tilde{\sigma} & \text{if } Q = R_\psi J\Sigma_3, \end{cases}$$

where $\tilde{\sigma} = (R_{-\phi}, 1)$.

Proof. The lemma follows from a computation which uses the commutation relations among SO(2), J, and Σ_3: $R_\psi J = JR_{-\psi}$, $R_\psi\Sigma_3 = \Sigma_3 R_\psi$, and $J\Sigma_3 = \Sigma_3 J$. ∎

Finally Lemma 4.22 allows us to decompose the load space $\mathcal{J}$ in a manner which facilitates the analysis of the bifurcation problem.

4.30 Lemma. *Let l be given by (3.32). Let*

$$\left.\begin{aligned} span\xi_3 &= \{\ p\xi_3\ |\ p \in \mathbb{R}\ \} \\ \mathcal{K} &= \{\ \bar{\xi} \in \mathcal{J}\ |\ k(\bar{\xi}, \gamma_0): e_3 \otimes e_3 = 0\ \} \end{aligned}\right\} \qquad (4.30)$$

a) With respect to the inner product on $\mathcal{J}$ *(4.30) specifies closed subspaces which decompose* $\mathcal{J}$ *as*

$$\mathcal{J} = span\xi_3 \oplus \mathcal{K}\ . \qquad (4.31)$$

Moreover, G *acts invariantly and orthogonally on the subspaces.*

b) For $(p, \bar{\xi}) \in \mathbb{R} \times \mathcal{K}$, *define* $m(p, \bar{\xi}) \in \mathcal{J}$ *by*

$$m(\mathrm{p},\bar{\xi}) = \mathrm{p}\xi_3 + \bar{\xi} \ . \tag{4.32}$$

Then m is an isometric isomorphism between the spaces. Moreover,

$$m(\mathrm{p},g\bar{\xi}) = gm(\mathrm{p},\bar{\xi}) \ . \tag{4.33}$$

Proof. By the definition of ξ_3 and by (3.47), for $0 \le \ell \le r-2$,

$$(\xi,\xi_3)_{\ell,2} = (\xi,\xi_3)_{0,2} = k(\bar{\xi},\gamma_0){:}\,e_3 \otimes e_3. \tag{4.34}$$

Hence, $\mathcal{K}$ is the subspace of elements of $\mathcal{J}$ orthogonal to spanξ_3 in the inner product of $\mathcal{J}$. By the theory of Hilbert spaces, the decomposition (4.31) follows. By (3.48), the definition of $\mathcal{K}$, and (4.22), G acts invariantly and orthogonally on the subspaces. By (4.34), the map $\mathrm{p} \longrightarrow \mathrm{p}\xi_3$ is an isometry between $\mathbb{R}$ and spanξ_3. Hence (4.31) implies that m is an isometric isomorphism. Finally, (4.33) follows from the invariance of (4.31) under the action of G. ∎

Problem 4.25 is an infinite-dimensional bifurcation problem, in that the sets of critical points we seek lie in the function manifold $\mathcal{M}_1$. In the next section we reduce the bifurcation problem to one on a finite-dimensional space.

V. THE REDUCTION OF THE BIFURCATION PROBLEM

In this section we combine the Liapunov-Schmidt reduction procedure used in [1-3] with the physically acceptable hypotheses on the response of the material comprising the rod made in [6] to reduce Problem 4.25 to a finite-dimensional bifurcation problem on a group. The reduction is taken relative to the entire orbit for the trivially equilibrating configuration. This reduction stands in contrast to those described in [31] and [34] which involve reductions at isolated singular points. Our analysis is more involved, because our unperturbed problem admits a continuous group of symmetries, rather than a discrete one.

V.1. *The Decomposition of the Spaces*

In this subsection we introduce the decompositions of the representation spaces for the virtual displacements and the applied loads which allow the reduction. The reduction then follows from a suitable hypothesis on the material response.

We begin by decomposing the representative spaces in a manner compatable with the constraints on the configurations and with the orbit of equilibrating configurations about which we are working (see [1] and [4]).

5.1 Lemma. For $\hat{\underline{e}}_j \in skew(E^3)$, $j = 1..3$, *let* $\hat{\underline{e}}_j \in \mathfrak{E}$ *denote the constant map* $\hat{\underline{e}}_j(S) \equiv \hat{\underline{e}}_j$. *Set*

$$o(2) = span_{\mathbb{R}}\{\hat{\underline{e}}_3\} \subseteq \mathfrak{E} .$$

$$o(2)^{\perp} = span_{\mathbb{R}}\{ \hat{\underline{e}}_\alpha \mid \alpha = 1, 2 \} \subseteq \mathfrak{E} .$$

a) $o(2)$ *and* $o(2)^{\perp}$ *are subspaces of* $\mathfrak{E}$ *invariant under the action of* Γ *of (4.27).*

b) Let $G\gamma_0$ *be given by (4.23). Let* G_1 *be given by (4.26). Then*

1) $G\gamma_0 = G_1\gamma_0$.

2) $T_{\gamma_0} G\gamma_0 = T_{\gamma_0} G_1\gamma_0 = o(2) \subseteq \mathfrak{C}$.

Proof. Part a) follows from (4.26) and Lemma 2.6. By Proposition 4.27 the mapping $g \longmapsto g\gamma_0$ of G onto $G\gamma_0$ is an embedding which factors through $G/\hat{\Gamma} \approx G_1$. Hence, using (2.21), b) follows. ∎

From Lemma 2.6, (4.22), and (4.29), it follows that the action of $\hat{\Gamma}$ on $o(2) \oplus o(2)^{\perp}$ in $\mathfrak{C}$, is determined by the usual action of Γ on $\mathbb{R}^3$. Its action on $o(2)$ is determined by that of Z_2 on $\mathbb{R}$, with $SO(2)$ acting trivially. Its action on $o(2)^{\perp}$ is determined by the usual action of $SO(2)$ on $\mathbb{R}^2$, and with $J\Sigma_3$ acting as the reflection element.

We will use two decompositions of $T_{\gamma_0} M$ which are consistent with the constraint.

5.2 Lemma. *Let* $\gamma_0 \in M_1$ *be given by Lemma 4.23.* *Set* $S_\beta(\gamma_0) = S_{\beta\gamma_0}$, $\beta = 1,2$ *be given by (4.17).* *Let* $\alpha = G_1\gamma_0$. *Let* B *be given by (4.16).* *Set*

$$N^{\alpha}_{\gamma_0}(M) = \{ Z \in \mathfrak{C} \mid B(Z,\hat{\underline{e}}_3) = 0 \}$$

$$N^{\alpha}_{\gamma_0}(M_1) = \{ Z \in T_{\gamma_0} M_1 \mid B(Z,\hat{\underline{e}}_3) = 0 \}$$

denote the orthogonal complements to $T_{\gamma_0} G\gamma_0$ *in* $T_{\gamma_0} M$ *and* $T_{\gamma_0} M_1$, *respectively.* *Let*

$$U_2(\gamma_0) = \{ Z \in \mathfrak{C} \mid B(Z,\hat{\underline{e}}_j) = 0, \ j = 1..3 \} \tag{5.1}$$

Then the following sums are direct and invariant under $\hat{\Gamma}$:

a) $T_{\gamma_0} M = o(2)^{\perp} \oplus T_{\gamma_0} G_1\gamma_0 \oplus U_2(\gamma_0)$

b) $N^{\alpha}_{\gamma_0}(M) = o(2)^{\perp} \oplus U_2(\gamma_0)$ $\tag{5.2}$

$N^{\alpha}_{\gamma_0}(M_1) = U_2(\gamma_0)$

$$T_{\gamma_0}\mathcal{M}_1 = S_2(\gamma_0) = T_{\gamma_0}G_1\gamma_0 \oplus U_2(\gamma_0) \tag{5.3}$$

Moreover, all sums are orthogonal with respect to B.

Proof. When $\gamma = \gamma_0$, $S_1(\gamma_0) = o(2)$ and $S_2(\gamma_0) = U_2(\gamma_0)$. So (4.17), (4.18), the definition for $U_2(\gamma_0)$, and the relation between B and K expressed in the proof of Proposition 4.18 gives the orthogonal direct sums for $T_{\gamma_0}\mathcal{M}$ and $T_{\gamma_0}\mathcal{M}_1$. Since $N^{\alpha}_{\gamma_0}(\mathcal{M})$ is an orthogonal complement, the orthogonal sum for $S_2(\gamma_0)$ gives $N^{\alpha}_{\gamma_0}(\mathcal{M}_1) = U_2(\gamma_0)$. By (2.23), (4.17), (4.18), and Proposition 4.27, the summands are invariant under the action of Γ. ∎

In an analogous manner we can decompose the representation space for the loads for the unconstrained problem.

5.3 Lemma. *For* $\hat{\underline{e}}_j \in skew(E^3)$, $j = 1..3$, *define* $\ell(\hat{\underline{e}}_j) \in \mathcal{F}$ *by*

$$\ell(\hat{\underline{e}}_j) = (\hat{q}_j, 0),\ \hat{q}_j(S) \equiv \hat{\underline{e}}_j.$$

Define the subspaces

$$\ell(o(2)^{\perp}) = span_{\mathbb{R}}\{\ \ell(\hat{\underline{e}}_{\alpha})\ \mid\ \alpha = 1,\ 2\ \} \subseteq \mathcal{F}$$
$$\ell(o(2)) = span_{\mathbb{R}}\{\ \ell(\hat{\underline{e}}_3)\ \} \subseteq \mathcal{F}$$
$$\mathcal{F}_{2e}(\gamma_0) = \{\ \zeta \in \mathcal{F}\ \mid\ K(\zeta, \hat{\underline{e}}_j) = 0,\ j = 1..3\ \}.$$

Then

$$\mathcal{F} = \ell(o(2)^{\perp}) \oplus \ell(o(2)) \oplus \mathcal{F}_{2e}(\gamma_0),$$

where the summands are invariant under the action of Γ *given by (3.44).*

Proof. The lemma follows from Lemma 4.19 and Proposition 4.27 in the manner of the proofs of Lemma 5.1a) and Lemma 5.2. ∎

We obtain the following relation between the decompositions.

5.4 Proposition. *Let K be the bilinear form of $\mathcal{F} \times \mathcal{C}$ given by Definition 3.31. Let*

$$\left.\begin{aligned} T_{\gamma_0}M &= o(2)^{\perp} \oplus T_{\gamma_0}G_1\gamma_0 \oplus U_2(\gamma_0) \\ \mathcal{F} &= j(o(2)^{\perp}) \oplus j(o(2)) \oplus \mathcal{F}_{2e}(\gamma_0) \end{aligned}\right\} \tag{5.4}$$

be the Γ invariant direct sums of Lemmas 5.2 and 5.3. Then

- *a) summands on the same vertical are in bijective correspondence under K,*
- *b) summands on different verticals are orthogonal with respect to K.*

In particular,

$$j(o(2)^{\perp}) \oplus j(o(2)) \text{ and } U_2(\gamma_0)$$

$$T_{\gamma_0}G_1\gamma_0 \oplus U_2(\gamma_0) \text{ and } \mathcal{F}_{2e}(\gamma_0)$$

are orthogonal complements under K.

Proof. The proposition follows from the nondegeneracy of K (Lemma 3.32), the relation between K and B given in the proof of Proposition 4.18, and Lemmas 5.2 and 5.3. ∎

To use (5.4) to reduce Problem 4.24 to a finite dimensional bifurcation problem we need to compute the linearization of the Euler operator for the rod (4.12) about γ_0 and determine how it acts on the subspaces of the decompositions. We obtain the following results.

5.5 Lemma. *For $p \in \mathbb{R}$, let $\xi_0 = \xi(p,0) \in \mathcal{J}$. Let $\Psi(\xi,\gamma)$ be the Euler operator (4.12). Let γ_0 be as given in Lemma 4.23. For $Z = \mathcal{C}$, define $L_pZ \in \mathcal{F}$ by*

$$L_pZ = [T_2\Psi(\xi_0,\gamma_0)]Z. \tag{5.5}$$

Then for $Z_1 = \hat{Z}_1$, $Z_2 = \hat{Z}_2$,

$$\left.\begin{aligned} &K(L_p Z_1, Z_2) = \\ &\int_I \{ \left[\frac{\partial m}{\partial u}(S,0)\right] z_1'(S)\cdot z_2'(S) - p(z_1(S) \times \underline{e}_3)\cdot(z_2(S) \times \underline{e}_3)\} dS, \end{aligned}\right\} \quad (5.6)$$

where $\frac{\partial m}{\partial u}(S,0) \in L(E^3)$ *is given by*

$$\left.\begin{aligned} &\frac{\partial m}{\partial u}(S,0) = \\ &2\frac{\partial H}{\partial \tau_1}(S,\underline{0})(\underline{e}_1 \otimes \underline{e}_1 + \underline{e}_2 \otimes \underline{e}_2) + 2\frac{\partial H}{\partial \tau_2}(S,\underline{0})(\underline{e}_3 \otimes \underline{e}_3), \end{aligned}\right\} \quad (5.7)$$

for $m(S,u)$ *and* $H(S,\tau)$ *as specified in Definition 4.10.*

Proof. By Lemma 2.6, for $Z_2 \in W^{r,2}(I,\text{skew}(E^3))$, $Z_2 = \hat{z}_2$ for some $\hat{z}_2 \in W^{r,2}(I,E^3)$. So by Lemma 2.6 and the proof of Proposition 4.11,

$$K(\Phi(\gamma),Z_2) = \int_I m(S,u(\gamma)(S))\cdot(z_2'(S) + u(\gamma)(S) \times z_2(S))dS.$$

By Lemma 4.9, if $H_\gamma(S) \in T_\gamma\mathcal{M}$, with $H_\gamma(S) = \gamma(S)\hat{z}(S)$,

$$([Tu(\gamma)]H_\gamma)(S) = z'(S) + u(\gamma)(S) \times z(S).$$

By the chain rule and Lemma 2.6,

$$\begin{aligned} &K([T\Phi(\gamma)]H_\gamma, Z_2) = \\ &\int_I \left[\frac{\partial m}{\partial u}(S,u(\gamma)(S))\right](z'(S) + u(\gamma)(S) \times z(S))\cdot(z_2'(S) + u(\gamma)(S) \times z_2(S))dS \\ &\quad + \int_I [m(S,u(\gamma)(S))\cdot(z'(S) + u(\gamma)(S) \times z(S)) \times z_2(S)]dS. \end{aligned}$$

Since $u(\gamma_0) \equiv 0$, (4.9) implies

$$K([T\Phi(\gamma_0)]H_\gamma, Z_2) = \int_I \left[\frac{\partial m}{\partial u}(S,0)\right] z'(S)\cdot z_2'(S)\,dS. \tag{5.8}$$

By Definition 3.31, for $H_\gamma(S) = \gamma(S)Z(S)$,

$$K([T_2\beta(\xi,\gamma)]H_\gamma, Z_2) = -\int_I \text{skew}(Z(S)\gamma^T(S)\mu(S)):Z_2(S)\,dS$$

$$-\text{skew}(Z(1)\gamma^T(1)\nu(1)):Z_2(1) - \text{skew}(Z(0)\gamma^T(0)\nu(0)):Z_2(0).$$

Taking $\gamma = \gamma_0$, $\xi = \xi_0$, $Z(S) = \hat{z}(S)$, $Z_2(S) = \hat{z}_2(S)$, and using Lemma 2.6,

$$\left.\begin{aligned} K([T_2\beta(\xi_0,\gamma_0]Z, Z_2) &= p\int_I \text{sk}(Z(S)\underline{e}_3 \otimes \underline{e}_3):Z_2(S)\,dS \\ = p\int_I Z(S)\underline{e}_3\cdot Z_2(S)\underline{e}_3\,dS &= p\int_I (z(S) \times \underline{e}_3)\cdot(z_2(S) \times \underline{e}_3)\,dS. \end{aligned}\right\} \tag{5.9}$$

Since $\Psi(\xi,\gamma) = \Phi(\gamma) - \beta(\xi,\gamma)$, (5.6) follows from (5.8) and (5.9). Equation (5.7) follows from the definition of $m(S,u(\gamma)(S))$. ∎

The symmetry of Ψ allows us to relate the linearization of the Euler operator about any configuration on the trivially equilibrating orbit $G\gamma_0$ to L_p.

5.6 Lemma. *If $g \in G$, if $Z \in \mathcal{E}$, then*

$$[T_2\Psi(\xi_0, g\gamma_0)]g\cdot Z = \mathcal{A}(g)L_pZ. \tag{5.10}$$

Proof. The lemma follows from (4.14), and (4.20). ∎

Following S. Antman ([9], p.296) we impose the following condition on the linear response of the material comprising the rod about the trivial configuration.

5.7 Hypothesis. *For $S \in I$, for $u \in E^3$, require*

$$\frac{\partial m}{\partial u}(S,u) \in L(E^3)$$

be positive definite on a neighborhood of $u = 0$. ∎

By equation (5.7) the hypothesis is equivalent to requiring that there be constants $c_2 > c_1 > 0$ for which

$$c_1 \le \frac{\partial H}{\partial \tau_1}(S,\tau) \le c_2 \text{ and } c_1 \le \frac{\partial H}{\partial \tau_2}(S,\tau) \le c_2. \tag{5.11}$$

on a neighborhood of $\tau = \underline{0}$ in R^2.

The hypothesis has a physical interpretation (see [19]). When the rod is in a configuration sufficiently near the initially straight configuration γ_0 in the C^1 sense, the hypothesis requires the material comprising the rod to oppose further bending and twisting about its centerline. The hypothesis of [9] is stronger than that presented here. It requires that Hypothesis 5.7 hold for all $u \in E^3$, or that the rod oppose further bending and twisting about its centerline, regardless of its configuration.

The hypothesis and the theory of elliptic differential equations give us a range of values for p over which we may use the decompositions (5.4). In the following proposition $W^{1,2}(I)$ is the Hilbert space of real-valued functions which are absolutely continuous on $I = [0,1]$, and which have weak first derivatives in $L^2(I)$.

5.8 Proposition. *For $p \in R$, let $L_p\colon T_{\gamma_0}\mathcal{M} \longrightarrow \mathcal{F}$ be as specified in Lemma 5.5. Set*

$$p_0 = \inf_{\substack{f \in V^1 \\ f \neq 0}} \frac{b(f,f)}{\|f\|^2_{L^2}}, \tag{5.12}$$

for $$V^1 = \{f \in W^{1,2}(I) \mid \int_I f(S)dS = 0\},$$

and $$b(f,f) = 2\int_I \frac{\partial H}{\partial \tau_1}(S,\underline{0})f'(S)f'(S)dS. \tag{5.13}$$

Let Hypothesis 5.7 hold. Then

a) $p_0 > 0$.

b) *For* $p < p_0$,

(1) $ker(L_p) = T_{\gamma_0}G_1\gamma_0$,

(2) $range(L_p) = \mathcal{F}_{2e}(\gamma_0) \oplus \mathfrak{j}(o(2)^\perp)$,

(3) L_p *determines isomorphisms between* $U_2(\gamma_0)$ *and* $\mathcal{F}_{2e}(\gamma_0)$, *and between* $o(2)^\perp$ *and* $\mathfrak{j}(o(2)^\perp$.

Proof.

a) *Lemma.* If $f \in V^1$, then $\| f' \|_{L^2} \geq \| f \|_{L^2}$.

Proof of the lemma. Since $W^{1,2}(I) \subseteq C^0(I)$, $f \in V^1$ and the Mean Value Theorem imply that

$$f(S) = \int_{s_0}^{s} f'(t)dt,$$

for some $s_0 \in I$. Thus $| f(S) | \leq \| f' \|_{L^2}$, giving the lemma. □

Part a) then follows from the lemma and (5.11).

b) Computations using (5.6) show that $K(L_p\hat{\underline{e}}_3,Z) = 0$ for all $Z \in \mathcal{E}$, that $K(L_p\hat{\underline{e}}_\alpha,Z) = 0$ for all $Z \in U_2(\gamma_0) \oplus o(2)$, that $K(L_pZ,\hat{\underline{e}}_j) = 0$ for all $Z \in U_2(\gamma_0)$, and that $K(L_p\hat{\underline{e}}_\alpha,\hat{\underline{e}}_\beta) = -p\delta_{\alpha\beta}$, for α, β = 1, 2, and j = 1, 2, 3. Hence (5.2) through (5.4) imply that $T_{\gamma_0}G_1\gamma_0 \subseteq \ker L_p$, $L(U_2(\gamma_0)) \subseteq \mathcal{F}_{2e}(\gamma_0)$, $L_p(o(2)^\perp) \subseteq \mathfrak{j}(o(2)^\perp)$, and that the restriction of L_p to $o(2)^\perp$ determines an isomorphism between $o(2)^\perp$ and $\mathfrak{j}(o(2)^\perp)$. It remains to show that the former two inclusions are onto, and that (5.12) holds. From (5.6) it follows that L_{p_1} is self-adjoint with respect to K. Hence, for $Z \in \mathcal{E}$, $Z_2 \in o(2)^\perp \oplus T_{\gamma_0}G_1\gamma_0$, $K(L_pZ,Z_2) = 0$. Proposition 5.4 then implies $L_pZ \in \mathcal{F}_{2e}(\gamma_0) = \mathcal{F}_e$. Let

$$U^1 = \{Z \in W^{1,2}(I,\text{skew}(E^3)) \mid \int_I Z(S)dS = 0\}.$$

From the lemma of part a),

$$(Z_1,Z_2)_1 \equiv (Z_1',Z_2')_{L^2}$$

is an inner product on U^1 whose norm is equivalent to that of $W^{1,2}$. Hence, U^1 is a Hilbert space under this norm. Moreover,

$$U_{\gamma_0} = U^1 \cap W^{r,2}(I,\mathrm{skew}(E^3)),$$

and the inclusion of U_{γ_0} into U^1 is a compact, dense injection. For Z_1, $Z_2 \in U^1$, for $p \in R$, define the bilinear form

$$a_p(Z_1,Z_2) = K(L_p Z_1, Z_2), \tag{5.14}$$

where the right hand side is given by (5.6). Since L_p is self-adjoint with respect to K, a_p is symmetric. By (5.11), a_0 is positive definite; hence, it is a coercive bilinear form with coercive constant $\lambda_0 = 0$ (see [20], p. 103). Since

$$a_p(Z_1,Z_2) = a_0(Z_1,Z_2) - p \int_I \mathcal{Z}_{1\alpha} \mathcal{Z}_{2\alpha}\, dS,$$

for $\mathcal{Z}_\alpha = Z:\hat{\underline{e}}_\alpha$, $\alpha = 1, 2$, using the theory of symmetric coercive forms in the manner of [20], p. 118 implies that (1) L_p determines a continuous, linear mapping $\tilde{L}_p$ of U^1 into its dual space which is surjective if and only if $\ker(\tilde{L}_p)$ is trivial; (2) $\ker(\tilde{L}_p)$ is finite dimensional for all p; and (3) $\ker(\tilde{L}_p)$ is not trivial for at most a countable collection of (positive) numbers with no finite accumulation point. Let p_1 be the smallest of these numbers. Then for $p < p_1$, for $\zeta \in \mathcal{F}_e$, (1) implies that there is a unique $Z \in U^1$ for which $K(L_p Z, Z_2) = K(\zeta, Z_2)$ for all $Z_2 \in U^1$. By standard regularity arguments (see [21] chap. 26), if $\zeta \in \mathcal{F}_e$, then $Z \in U^1 \cap W^{r,2} = U_2(\gamma_0)$. So $\mathcal{F}_e \subseteq \tilde{L}_p(U_{\gamma_0})$, and $\tilde{L}_p$ is an isomorphism between $U_2(\gamma_0)$ and $\mathcal{F}_{2e}(\gamma_0)$. The decomposition (5.4) implies that $T_{\gamma_0} G_1 \gamma_0 = \ker L_p$ and that $\mathrm{range}(L_p) = \mathcal{F}_{2e}(\gamma_0)$.

We now establish (5.12). For $f \in V^1$, $p \in R$, define $\ell_p(f)$ to be the linear function on V^1 given by

$$\langle \ell_p(f), g\rangle = b(f,g) - p(f,g)_{L^2} \equiv b_p(f,g), \tag{5.15}$$

for $g \in V^1$. By (5.11), b is a coercive bilinear form with coercive constant $\lambda_0 = 0$. By [20], p. 118, the three properties deduced above for $\tilde{L}_p$ also hold for ℓ_p. In the manner of [22], p. 337f, it follows that p_0 given by (5.12) is the smallest number for which $\ker(\ell_p)$ is not trivial. We now show that $p_1 = p_0$. Suppose that $Z \in \ker(\tilde{L}_{p_1})$. Taking $Z_2 = (Z:\hat{\underline{e}}_3)\hat{\underline{e}}_3$ and using (5.6) implies that $Z:\hat{\underline{e}}_3 = 0$. So for $Z_2 \in U_2(\gamma_0)$ arbitrary, (5.6), (5.13), and (5.15) imply

$$0 = K(L_{p_1}Z, Z_2) = b_{p_1}(\mathcal{Z}_1, \mathcal{Z}_{2_1}) + b_{p_1}(\mathcal{Z}_2, \mathcal{Z}_{2_2}),$$

where $\mathcal{Z}_\alpha = Z:\hat{\underline{e}}_\alpha \in V^1 \cap W^{r,2}$, $\alpha = 1, 2$. Since $V^1 \cap W^{r,2}(I)$ is dense in V^1 and b is positive definite, $\mathcal{Z}_\alpha \in \ker(\ell_{p_1})$, $\alpha = 1, 2$. Hence, $\ker(\ell_{p_1})$ is not trivial, implying $p_1 \geq p_0$. Conversely, if $f \in \ker(\ell_{p_0})$, then $f \in V^1 \cap W^{r,2}(I)$ by regularity arguments. Set $Z = f\hat{\underline{e}}_1$. Then (5.6) implies that $Z \in U_{\gamma_0} \cap \ker(L_{p_0})$. So $\ker(\tilde{L}_{p_0})$ is not trivial; hence $p_0 \geq p_1$. So $p_1 = p_0$. ∎

Combining these results, we obtain the theorem which will allow us to reduce Problem 4.25 to a finite dimensional bifurcation problem over the range $p < p_0$. (Compare to [3], p. 218 and [4], p. 322.)

5.9 Theorem. *Let $p \in \mathbb{R}$. Let $L_p : T_{\gamma_0}\mathcal{M} \longrightarrow \mathcal{F}$ be defined by (5.5). Let Hypothesis 5.7 hold. Then*

$$T_{\gamma_0}\mathcal{M} = o(2)^\perp \oplus T_{\gamma_0}G_1\gamma_0 \oplus U_2(\gamma_0)$$

$$\mathcal{F} = \mathcal{j}(o(2)^\perp) \oplus \mathcal{j}(o(2)) \oplus \mathcal{F}_{2e}(\gamma_0)$$

and the direct sums are invariant under the actions of Γ.

a) Relative to K, the summands on the same verticals are in bijective correspondence, and those off a given vertical are orthogonal.

b) Let p_0 be defined by Proposition 5.8. Then for $p < p_0$, $\mathrm{range}(L_p) = \mathcal{F}_{2e}(\gamma_0) \oplus \mathcal{j}(o(2)^\perp)$, $\ker L_p = T_{\gamma_0}G_1\gamma_0$, and the *restrictions of L_p to $o(2)^\perp$ and to $U_2(\gamma_0)$ determine isomorphisms onto $j(o(2)^\perp$ and $\mathcal{F}_{2e}(\gamma_0)$, respectively.*

Proof. The theorem restates the conclusions of Propositions 5.4 and 5.8. ∎

The number p_0 is significant. As indicated in the proof of Proposition 5.8, it is the smallest value for p for which the Sturm-Liouville problem

$$\left.\begin{aligned} &2\left(\frac{\partial H}{\partial \tau_1}(S,\underline{0})f'(S))\right)' + pf(S) = 0, \\ &f'(1) = f'(0) = 0, \\ &\int_I f(S)dS = 0 \end{aligned}\right\} \tag{5.16}$$

has non-trivial solution. By the relation of $\ker(L_p)$ and $\ker(\ell_p)$ deduced in the proof of Proposition 5.8, we may physically interpret p_0 as the smallest pressure on the rod for which buckling will commence.

When $p = p_0$ the decompositions of (5.4) are inadequate to reduce Problem 4.25 to a finite dimensional bifurcation problem. Rather, a finer decomposition is needed, depending upon the nature of p_0. If p_0 is a simple eigenvalue for (5.16), a generic assumption (see [22], p. 337), we have the following result (see [4], p. 331).

5.10 Theorem. *Let p_0 be a simple eigenvalue for (5.16). Let f_0 be a normalized eigenfunction associated with p_0. Let L_{p_0} be defined by (5.5). Let Hypothesis 5.7 hold. Define*

$$W_3(\gamma_0) = \operatorname{span}_{\mathbb{R}}\{\ f_0\hat{\underline{e}}_\alpha \mid \alpha = 1,\ 2\ \} \subseteq U_2(\gamma_0),$$

$$\mathcal{F}_{3e}(\gamma_0) = \{\zeta \in \mathcal{F} \mid \zeta = (\hat{q},0),\ \hat{q}(S) = f_0(S)c_\alpha\hat{\underline{e}}_\alpha.\ c_\alpha \in \mathbb{R}\} \subseteq \mathcal{F}_e,$$

$$W_4(\gamma_0) = \{Z \in U_2(\gamma_0) \mid K(\zeta,Z) = 0,\ \zeta \in \mathcal{F}_{3e}(\gamma_0)\},$$

$$\mathcal{F}_{4e}(\gamma_0) = \{\zeta \in \mathcal{F} \mid K(\zeta,Z) = 0,\ Z \in W_3(\gamma_0)\}.$$

Then

a) $$T_{\gamma_0}\mathcal{M} = o(2)^{\perp} \oplus T_{\gamma_0}G_1\gamma_0 \oplus W_3(\gamma_0) \oplus W_4(\gamma_0), \tag{5.17}$$

$$\mathcal{F} = j(o(2)^{\perp}) \oplus j(o(2)) \oplus \mathcal{F}_{3e}(\gamma_0) \oplus \mathcal{F}_{4e}(\gamma_0), \tag{5.18}$$

and the direct sums are invariant under the action of Γ.

b) *Relative to* K, *summands on the same vertical are in bijective correspondence, summands on different verticals are orthogonal.*

c) $KerL_{p_0} = T_{\gamma_0}G_1\gamma_0 \oplus W_3(\gamma_0)$,

$Range(L_{p_0}) = \mathcal{F}_{4e}(\gamma_0) \oplus j(o(2)^{\perp}),$

and the restriction of L_{p_0} *to* $o(2)^{\perp}$ *and to* $W_4(\gamma_0)$ *determine isomorphisms onto* $j(o(2)^{\perp})$ *and* $\mathcal{F}_{4e}(\gamma_0)$, *respectively.*

Proof. The theorem follows in the manner of Proposition 5.4 and 5.8. ∎

V.2 *The Liapunov-Schmidt Reduction*

The decompositions of Theorem 5.9 and 5.10 allow us to use a Liapunov-Schmidt reduction procedure to reduce Problem 4.25 to a finite-dimensional bifurcation problem on the group G_1. Moreover, the reduction is accomplished in such a way that the group Γ continues to act as a group of symmetries in the finite-dimensional problem. In this subsection we carry out in detail the reduction for the case where $p < p_0$ (Theorem 5.18). We then indicate how to modify the reduction when $p = p_0$ (Theorem 5.24).

We begin by introducing a tubular neighborhood for the orbit $G\gamma_0$ of trivially equilibrating configurations which arises in $\mathcal{M}$ from the Lie group structure it inherits from $O(E^3)$.

5.11 Lemma. *Let $N^{\alpha}_{\gamma_0}(\mathcal{M})$ be given by Lemma 5.2. There is a neighborhood N_ε of 0 in $N^{\alpha}_{\gamma_0}(\mathcal{M})$ and a smooth map ρ of $G_1 \times N_\varepsilon$ into $\mathcal{M}$ such that*

a) $\rho(G_1 \times N_\varepsilon)$ is an open neighborhood of $G_1\gamma_0$ in $\mathcal{M}$, and ρ is a diffeomorphism onto it;

b) for $\sigma \in G_1$, $g \in \Gamma$, $Z \in N_\varepsilon$,

$$(1) \quad \rho(\sigma,0) = \mathcal{T}_\sigma \gamma_0,$$

$$(2) \quad \rho(\sigma,Z) = \mathcal{T}_\sigma \rho(1,Z),$$

$$(3) \quad \rho(Ad_g\sigma, \mathcal{B}(g)Z) = \mathcal{T}_g \rho(\sigma,Z),$$

$$(4) \quad T_2\rho(\sigma,0) = T\mathcal{T}_\sigma(\gamma_0).$$

Here, $\mathcal{B}(g)$ is given by (2.23), $Ad_g\sigma$ by (4.29), and $\mathcal{T}_g$ denotes the action of G on $\mathcal{M}$ given by (2.9).

Proof. Let $\exp_1 : T_1O(E^3) \longrightarrow O(E^3)$ be the exponential map on $O(E^3)$ based at the unit element ([7], chapter 4). Recall that, for $\exp_Q : T_QO(E^3) \longrightarrow O(E^3)$, for $W \in T_1O(E^3)$, $\exp_Q(TL_Q(1)W) = L_Q\exp_1(W)$. If $Z \in N^{\alpha}_{\gamma_0}(\mathcal{M})$, let $\mathit{exp}(Z) \in \mathcal{M}$ be the induced map

$$(\mathit{exp}(Z))(S) = \exp_1(Z(S)) \in O(E^3) .$$

Set $\rho(\sigma,Z) = \mathcal{T}_\sigma \mathit{exp}(Z)$. Then ρ satisfies properties (1), (2), and (4) of b), and by Elliason [28] is a local diffeomorphism of $G_1 \times N_\varepsilon$ onto a neighborhood of $G_1\gamma_0$. Finally, property (3) follows from property (2) and the fact that $\exp_1(QWQ^T) = Q\exp_1(W)Q^T$ in $O(E^3)$. ∎

Of particular value for the future analysis is the following lemma concerning the derivative of the tubular neighborhood map. It is convenient to introduce it at this time. This property also arises from the Lie group structure induced on $\mathcal{M}$ by $O(E^3)$.

5.12 Lemma. *If $Z \in N_\varepsilon$, $W \in C$, and if $\rho(1,Z)$ is given by Lemma 5.11, then*

$$[T_2\rho(1,Z)]W = T\mathcal{L}_{\rho(1,Z)}(\gamma_0)M_Z W \in T_{\rho(1,Z)}\mathcal{M} , \tag{5.19}$$

Where $M_Z W \in \mathcal{C}$ is given by

$$M_Z W = \sum_0^\infty \frac{1}{(n+1)!}\left(-adZ\right)^n W , \tag{5.20}$$

a uniformly convergent series on I, with $(adZ)W = [Z,W]$.

Proof. From [29] p.95, if $X, Y \in T_1 O(E^3) = o(E^3)$, $[Texp_1(X)]Y = TL_X(1)M_X Y$, where $M_X Y \in o(E^3)$ is given by

$$M_X Y = \sum_0^\infty \frac{1}{(n+1)!}\left(-adX\right)^n Y ,$$

a convergent series. Since

$$([T\mathit{exp}(Z)]W)(S) = [Texp_1(Z(S))](W(S)),$$

arguing in the manner of Elliason [28] gives

$$([T\mathit{exp}(Z)]W)(S) = \left((M_{Z(S)}W(S)\right)$$

pointwise on I. The differentiability of the "ad" map on $\mathcal{C}$ (see [7] and [28]), the compactness of I, and Lemma 2.18 imply that (5.20) converges uniformly on I, $M_Z W \in \mathcal{C}$, and that in fact M_Z is a smooth map on $\mathcal{C}$. Since $\rho(1,Z) = \mathit{exp}(Z)$, Lemma 5.11 then gives the result. ∎

We extract a tubular neighborhood for $G_1\gamma_0$ for the constrained problem by identifying a submanifold of $G_1 \times N_\varepsilon$ which is taken by ρ into $\mathcal{M}_1$ locally about $G_1\gamma_0$.

5.13 Lemma. *Let $U(G\gamma_0) = \rho(G_1 \times N_\varepsilon)$ be given by Lemma 5.11. Let i be given by Lemma 4.16. For $(\sigma,Z) \in G_1 \times N_\varepsilon$, define $j(\sigma,Z) \in \mathbb{R}^2$ by*

$$j(\sigma,Z) = i(\rho(\sigma,Z)) . \tag{5.21}$$

Let $$\mathfrak{J}_1 = \{Z \in N_\varepsilon \mid j(1,Z) = 0\} .$$

a) Then

(1) $Z \in \mathfrak{J}_1$ implies $\rho(1,Z) \in \mathcal{M}_1$;

(2) $\mathfrak{J}_1$ is a smooth submanifold of N_ε about 0 in $N^\alpha_{\gamma_0}(\mathcal{M})$ of codimension 2.

b) If $Z \in \mathfrak{J}_1$, and if $\xi_\alpha \in \mathcal{G}$ is given by Lemma 4.16, then

$$T_Z\mathfrak{J}_1 = \{W \in N^\alpha_{\gamma_0}(\mathcal{M}) \mid K(\beta(\xi_\alpha,\rho(1,Z)),M_Z W) = 0\} ,$$

where $M_Z W$ is given by (5.20). In particular,

$$T_0\mathfrak{J}_1 = U_2(\gamma_0),$$

given by (5.1).

c) Let $Z \in \mathfrak{J}_1$ and $\hat{\underline{e}}_3 \in \mathfrak{C}$. Set

$$\langle Ad_{\rho^T(1,Z)}\hat{\underline{e}}_3\rangle = \text{span}_{\mathbb{R}}\{\rho(1,Z)\underline{e}_3\} \subseteq \mathfrak{C} .$$

Then

$$S_2(\rho(1,Z)) = M_Z T_Z\mathfrak{J}_1 \oplus \langle Ad_{\rho^T(1,Z)}\hat{\underline{e}}_3\rangle , \tag{5.22}$$

a direct sum, where $S_2(\gamma) = S_{2\gamma}$ is given by (4.18).

d) $(\mathfrak{J}_1)_\varepsilon = \mathfrak{J}_1 \cap N_\varepsilon$ is an open submanifold of $\mathfrak{J}_1$ about 0, and

$$\rho\colon\ G_1 \times (\mathfrak{J}_1)_\varepsilon \longrightarrow \mathcal{M}_1$$

is a local diffeomorphism about $G_1 \times 0$ which defines an open tubular neighborhood of $G\gamma_0$ in $\mathcal{M}_1$.

Proof.

a) Lemma 4.16 and (5.21) imply that $\rho(1,Z) \in \mathcal{M}_1$ when $Z \in \mathfrak{Z}_1$. From (5.21) and Lemma 4.16 $T_2 j(1,0)$ is a surjective map of $N^{\alpha}_{\gamma_0}(\mathcal{M})$ onto $\mathbb{R}^2$. So $0 \in \mathbb{R}^2$ is a regular value for $j(1,\cdot)$, and the Implicit Function Theorem implies part a).

b) The characterization of the tangent space follows from (4.15), (5.20), and (5.21). Since $M_0 W = W$, Lemma 5.11a) and (5.1) give the characterization of $T_0\mathfrak{Z}_1$.

c) By Lemma 4.22, $\mathcal{M}_1$ is invariant under G_1. So

$$T_{\rho(1,z)} G_1 \rho(1,z) \subseteq T_{\rho(1,z)} \mathcal{M}_1 \ .$$

A computation gives

$$T\mathcal{L}_{\rho(1,z)}(\gamma_0) \mathrm{Ad}_{\rho^T(1,z)} \hat{\underline{e}}_3 = T_{\rho(1,z)} G_1 \rho(1,z) \ .$$

By (4.18),

$$\langle Ad_{\rho^T(1,Z)} \hat{\underline{e}}_3 \rangle \subseteq S_2(\rho(1,Z)) \ .$$

By (4.17)

$$M_Z T_Z \mathfrak{Z}_1 \subseteq S_2(\rho(1,Z)) \ .$$

At $Z = 0$, Lemma 5.2 and part b) imply that

$$M_0 T_0 \mathfrak{Z}_1 \cap \langle Ad_{\rho^T(1,0)} \hat{\underline{e}}_3 \rangle = U_2(\gamma_0) \cap o(2) = 0 \ .$$

By transversality, for $Z \in \mathfrak{Z}_1$ sufficiently small, the summands involved in (5.22) have trivial intersection; hence, the sum is direct and contained in $S_2(\rho(1,Z))$. By Proposition 4.18,

$$\left.\begin{aligned} &S_1(\rho(1,Z)) \oplus (M_Z T_Z \mathfrak{Z}_1 \oplus \langle \mathrm{Ad}_{\rho^T(1,Z)} \hat{\underline{e}}_3 \rangle) \subseteq \\ &S_1(\rho(1,Z)) \oplus S_2(\rho(1,Z)) = T_{\gamma_0} \mathcal{M} \end{aligned}\right\} \qquad (5.23)$$

where the first sum in the first line is orthogonal with respect to B of (4.16). From Lemma 5.2, the codimension of $U_2(\gamma_0)$ in $T_{\gamma} \mathcal{M}$

is 3. By part a) and the definition of $N^{\alpha}_{\gamma_0}(\mathcal{M})$, $\mathfrak{Z}_1$ has the same codimension. So for $Z \in \mathfrak{Z}_1$ sufficiently small, $T_Z\mathfrak{Z}_1$ has codimension 3 in $T_{\gamma_0}\mathcal{M}$. Since $M_0 = 1_{\mathcal{E}}$ from (5.20), and since the set of isomorphisms of $\mathcal{E}$ is open in $L(\mathcal{E},\mathcal{E})$, M_Z is an isomorphism for Z sufficiently small. Hence, $M_Z T_Z\mathfrak{Z}_1$ has codimension 3 in $T_{\gamma_0}\mathcal{M}$. By construction,

$$S_1(\rho(1,Z)) \oplus \langle \mathrm{Ad}_{\rho^T(1,Z)} \hat{\underline{e}}_3 \rangle$$

is a codimension 3 subspace of $T_{\gamma_0}\mathcal{M}$. Hence, the inclusion in (5.23) is an equality, and the orthogonality of the first sum gives (5.22).

d) A computation from (4.17) shows that for $\sigma \in G_1$, $S_2(\mathcal{T}_\sigma \gamma) = S_2(\gamma)$. Since $\rho(\sigma,Z) = \mathcal{T}_\sigma \rho(1,Z)$ and G_1 is compact, we may choose $\varepsilon > 0$ sufficiently small that the family of manifolds

$$\{ \rho(\sigma,\mathfrak{Z}_{1\varepsilon}) \mid \sigma \in G_1 \}$$

do not intersect. With part c), part d) then follows. ∎

Using Lemma 5.13 we can rephrase Problem 4.25.

5.14 Lemma. *Let $\rho(G_1 \times \mathfrak{Z}_{1\varepsilon}) \equiv U_1(G\gamma_0) \subseteq \mathcal{M}_1$ be the neighborhood given by Lemma 5.13. Let V_1 be given by Problem 4.25. For $(\xi,\sigma,Z) \in \mathcal{J} \times G_1 \times \mathfrak{Z}_{1\varepsilon}$ define $V(\xi,\sigma,Z) \in \mathbb{R}$ by*

$$V(\xi,\sigma,Z) = V_1(\xi,\rho(\sigma,Z)) \; . \tag{5.24}$$

Then $\mathrm{crit} V_1(\xi,\cdot) \cap U_1(G\gamma_0)$ is in bijective correspondence with $\mathrm{crit}\, V(\xi,\cdot,\cdot)$ in $G_1 \times \mathfrak{Z}_{1\varepsilon}$.

Proof. Since ρ is a local diffeomorphism, the bijection follows using (5.24) and the chain rule. ∎

To facilitate our use of the decompositions of Theorem 5.9 and 5.10, it is convenient to describe $\mathfrak{Z}_{1\varepsilon}$ in terms of a fixed subspace of $N^{\alpha}_{\gamma_0}(\mathcal{M})$.

5.15 Lemma. *Let $U_{2\varepsilon} = N_\varepsilon \cap U_2(\gamma_0)$. Choosing ε sufficiently small, there is a smooth local diffeomorphism Ω of $U_{2\varepsilon}$ onto $\mathfrak{I}_{1\varepsilon}$ about 0 in N_ε such that*

$$\left.\begin{array}{ll} a) & \Omega(0) = 0, \\ b) & T\Omega(0) = 1 \in L(U_2(\gamma_0), U_2(\gamma_0)), \\ c) & \text{If } g \in \Gamma,\ \Omega(\mathcal{B}(g)Z) = \mathcal{B}(g)\Omega(Z)\ , \end{array}\right\} \quad (5.25)$$

where $\mathcal{B}(g)$ acts on $\mathfrak{E}$ by (2.23).

Proof. For $(Z_1,Z_2) \in o(2)^\perp \oplus U_{2\varepsilon}$ define $i_2(Z_1,Z_2) = j(1,Z_1+Z_2) \in \mathbb{R}^2$ using (5.2). By (4.15) and Lemma 5.11 $T_1i_2(0,0)$ is an isomorphism of $o(2)^\perp$ and $\mathbb{R}^2$. Since $j(1,0) = 0$, the Implicit Function Theorem implies that, choosing $\varepsilon > 0$ sufficiently small, there is a smooth map θ of $U_{2\varepsilon}$ into $o(2)^\perp$ such that $\theta(0) = 0$, and

$$i_2(Z_1,Z_2) = 0,\ Z_1+Z_2 \in N_\varepsilon, \text{ if and only if } Z_1 = \theta(Z_2).$$

Implicitly differentiating $i_2(\theta(Z_2),Z_2) = 0$ using Lemma 5.11 and (4.15) gives $T\theta(0) = 0 \in L(U_2(\gamma_0),o(2)^\perp)$. Setting $\Omega(Z_2) = Z_2 + \theta(Z_2)$ gives the desired map and the first two properties. The third property follows from Lemmas 5.2, 5.11, and 4.16. ∎

5.16 Proposition. *Let V be given by Lemma 5.14. For $(\xi,\sigma,Z) \in \mathcal{J} \times G_1 \times U_{2\varepsilon}$, define $V_1(\xi,\sigma,Z) \in \mathbb{R}$ by*

$$V_1(\xi,\sigma,Z) = V(\xi,\sigma,\Omega(Z))\ . \quad (5.26)$$

a) Then $\mathrm{crit}V_1(\xi,\cdot,\cdot)$ in $G_1 \times U_{2\varepsilon}$ is in bijective correspondence with $\mathrm{crit}V(\xi,\cdot,\cdot)$ in $G_1 \times \mathfrak{I}_{1\varepsilon}$.
b) $(\sigma,Z) \in \mathrm{crit}V_1(\xi,\cdot,\cdot)$ if and only if

$$1)\quad K\big(\Psi(\sigma^{-1}\xi,\rho(1,\Omega(Z))),M_{\Omega(Z)}T\Omega(Z)W\big) = 0 \quad (5.27)$$

for all $W \in U_2(\gamma_0)$,

$$2)\ \sigma \in \mathrm{crit}\ h(\xi,\cdot), \quad (5.28)$$

where $h(\xi,\cdot): G_1 \longrightarrow \mathbb{R}$ *is given by*

$$h(\xi,\sigma) = V_1(\xi,\sigma,Z).$$

Proof. Since Ω is a local diffeomorphism about 0 of $U_2(\gamma_0)$ and $\mathfrak{J}_{1\varepsilon}$ part a) follows from (5.26). By the chain rule, (4.11), (4.14), and (5.19), for $W \in U_2(\gamma_0)$,

$$\left[T(V_1(\xi,\cdot,\cdot))(\sigma,Z)\right]\left(TL_\sigma(1)\hat{\underline{e}}_3,W\right) =$$

$$T_2(V(\xi,\sigma,\Omega(Z)))TL_\sigma(1)\hat{\underline{e}}_3 + T_3V(\xi,\sigma,\Omega(Z))T\Omega(Z)W =$$

$$T(h(\xi,\cdot))(\sigma)TL_\sigma(1)\hat{\underline{e}}_3 + T_2V(\sigma^{-1}\xi,\rho(1,\Omega(Z)))T_2\rho(1,\Omega(Z))T\Omega(Z)W =$$

$$T(h(\xi,\cdot))(\sigma)TL_\sigma(1)\hat{\underline{e}}_3 + K\left(\Psi(\sigma^{-1}\xi,\rho(1,\Omega(Z))),M_{\Omega(Z)}T\Omega(Z)W\right) .$$

If (5.27) and (5.28) hold, the above equations imply $(\sigma,Z) \in \mathrm{crit}V_1(\xi,\cdot,\cdot)$. Conversely, taking in turn $W = 0$ and W arbitrary in the above equations give (5.27) and (5.28). ∎

By Theorem 5.9 we may satisfy the first condition of Proposition 5.16 by an implicitly defined function.

5.17 Proposition. *Let* $\mathcal{K}$ *be given by (4.30). Let* $\bar{p} \in (0,p_0)$. *If* $[p_1,p_2] \subseteq (0,p_0)$ *contains* $\bar{p}$ *in its interior, then there are neighborhoods* $\mathcal{B}_\eta$ *and* $U_{2\varepsilon}$ *about the zero elements in* $\mathcal{K}$ *and* $U_2(\gamma_0)$, *respectively, and a smooth map*

$$\psi : (p_1,p_2) \times \mathcal{B}_\eta \times G_1 \longrightarrow U_{2\varepsilon}$$

such that for $p \in (p_1,p_2)$, $\bar{\xi} \in \mathcal{B}_\eta$, $\xi = p\xi_3+\bar{\xi}$, *and* $Z \in U_{2\varepsilon}$, *(5.27) holds if and only if* $Z = \psi(p,\bar{\xi},\sigma)$.

a) The neighborhoods $\mathcal{B}_\eta$ *and* $U_{2\varepsilon}$ *are invariant under the actions of* G *and* Γ, *respectively, and* ψ *satisfies the following properties:*

$$\left.\begin{array}{l} (1)\ \ \psi(p,0,\sigma) \equiv 0 \ for\ \sigma \in G_1, \\ (2)\ \ \psi(p,\sigma_1\bar{\xi},\sigma_1\sigma) = \psi(p,\bar{\xi},\sigma)\ for\ \sigma_1 \in G_1, \\ (3)\ \ \psi(p,g\bar{\xi},Ad_g\sigma) = \mathcal{B}(g)\psi(p,\bar{\xi},\sigma)\ for\ g \in \Gamma, \end{array}\right\} \quad (5.29)$$

where $Ad_g\sigma \in G_1$ *is given by (4.29).*

b) If $[\bar{p}_1,\bar{p}_2] \subseteq (0,p_0)$ *is a second set containing* $\bar{p}$ *in its interior, and if* $\mathcal{B}_{\eta 2}$, $U_{2\varepsilon 2}$, *and* ψ_2 *are the neighborhoods and map associated with it in the manner of part a), then* ψ *and* ψ_2 *coincide on the intersections of their domains.*

Proof. By Theorem 5.9 we may identify the dual space of $U_2(\gamma_0)$ with $\mathcal{F}_{2e}(\gamma_0)$. For $(p,\bar{\xi},\sigma,Z) \in \mathbb{R} \times \mathcal{K} \times G_1 \times U_2(\gamma_0)$ and for $W \in U_2(\gamma_0)$, define $\Sigma(p,\bar{\xi},\sigma,Z) \in \mathcal{F}_{2e}(\gamma_0)$ by

$$K(\Sigma(p,\bar{\xi},\sigma,Z),W) = K\big(\Psi(\xi,\rho(\sigma,\Omega(Z))),M_{\Omega(Z)}T\Omega(Z)W\big) ,$$

where $\xi = p\xi_3 + \bar{\xi}$. By Lemmas 5.11, 5.12, and 4.23, for $p \in \mathbb{R}$, $\sigma \in G_1$ fixed but arbitrary, for $W,\ V \in U_2(\gamma_0)$,

$$K(\Sigma(p,0,\sigma,0),W) = 0,$$

$$K([T_4\Sigma(p,0,\sigma,0)]V,W) = K(L_pV,W) .$$

By Theorem 5.9 and the nondegeneracy of K,

$$\Sigma(p,0,\sigma,0) = 0 ,$$

$$T_4\Sigma(p,0,\sigma,0) = L_p .$$

If $\bar{p} < p_0$, $L_{\bar{p}}$ is an isomorphism of $U_2(\gamma_0)$ onto $\mathcal{F}_{2e}(\gamma_0)$. The Implicit Function Theorem implies there is a neighborhood $\mathcal{W} = \mathcal{I}_\alpha \times \mathcal{B}_\eta \times I_\delta$ about $(\bar{p},0,\sigma)$ in $(o,p_0) \times \mathcal{K} \times G_1$, a neighborhood $U_{2\varepsilon}$ of zero in $U_2(\gamma_0)$, and a smooth map $\psi_{\bar{p},\sigma}$ between them for which, if $(p,\bar{\xi},\bar{\sigma}) \in \mathcal{W}$ and $Z \in U_2(\gamma_0)$, $\Sigma(p,\bar{\xi},\bar{\sigma},Z) = 0$ if and only if $Z = \psi_{\bar{p},\sigma}(p,\bar{\xi},\bar{\sigma})$. Here, α, η, δ, and $\psi_{\bar{p},\sigma}$ depend upon the choice of $\bar{p}$ and σ. By their definitions the locally defined functions $\psi_{\bar{p},\sigma}$ coincide on the intersections of their domains. Hence, using the compactness of G_1 and a partition of unity, we can find neighborhoods $\mathcal{I}_\alpha$, $\mathcal{B}_\eta$, and $U_{2\varepsilon}$ about $\bar{p}$, 0, and 0, respectively, and a smooth function $\psi_{\bar{p}}$ taking $\mathcal{I}_\alpha \times \mathcal{B}_\eta \times G_1$ into $U_{2\varepsilon}$ having the desired property involving (5.27). Here, α, η, ε, and $\psi_{\bar{p}}$ depend upon the choice of $\bar{p}$. By construction, the $\psi_{\bar{p}}$ coincide on the intersections of their domains. Hence, using the compactness of $[p_1,p_2]$ and a partition of unity subordinate to a finite collection of the $\{\mathcal{I}_\alpha\}$ covering $[p_1,p_2]$, we can obtain neighborhoods $\mathcal{B}_\eta$ and $U_{2\varepsilon}$ and a smooth function ψ taking $(p_1,p_2) \times \mathcal{B}_\eta \times G_1$ into $U_{2\varepsilon}$ which depend only upon the choice of p_1 and p_2, and which have the desired property involving (5.27). The invariance of $\mathcal{B}_\eta$ and $U_{2\varepsilon}$ under the action of G and Γ, respectively, follows from Lemmas 2.22 and 3.33. Properties (5.29) follow from the uniqueness of ψ, Lemma 5.11, (4.14), (5.25), and the invariance of $\mathcal{F}_{2e}(\gamma_0)$ and $U_2(\gamma_0)$ under the action of Γ, established in Lemma 5.3. Hence part a) follows. Part b) follows from the manner in which ψ is constructed. ∎

Remark. Property (3) of (5.29) is noteworthy. The $\mathrm{Ad}_g\sigma$ term arises, because G is a *semidirect* product of G_1 and Γ, not a direct product. Specifically, $\mathrm{Ad}_g\sigma$ specifies the commutator between the elements g and σ of the two subgroups. Mathematically, it portrays the extent to which the product is not direct. Physically, it manifests itself as a coupling between rotations about the centerline of a rod and its spatial rotation about the axis on which its ends are constrained to lie.

We obtain the following reduction of Problem 4.25.

5.18 Theorem. *Let $\bar{p} < p_0$ and let $[p_1,p_2] \subseteq (0,p_0)$ be any compact neighborhood of $\bar{p}$. Let $\mathcal{B}_\eta$, $U_{2\varepsilon}$, ρ, and ψ be as specified in Lemmas 5.11 and 5.15, and Proposition 5.17. For $p \in (p_1,p_2)$, $\bar{\xi} \in B_\eta$, and $\sigma \in G_1$, define the smooth, real-valued function $f(p,\bar{\xi},\sigma)$ by*

$$f(p,\bar{\xi},\sigma) = V_1(\xi,\sigma,\psi(p,\bar{\xi},\sigma)) = V_1(\xi,\rho(\sigma,\Omega(\psi(p,\bar{\xi},\sigma)))), \qquad (5.30)$$

where $\xi = p\xi_3 + \bar{\xi}$.

a) Then $f(p,\bar{\xi},\sigma)$ satisfies the following properties:

$$\left.\begin{aligned}&(1)\quad f(p,0,\sigma) = f(p,0,1) \text{ for } \sigma \in G_1,\\ &(2)\quad f(p,\sigma_1\bar{\xi},\sigma_1\sigma) = f(p,\bar{\xi},\sigma) \text{ for } \sigma_1 \in G_1,\\ &(3)\quad f(p,g\bar{\xi},Ad_g\sigma) = f(p,\bar{\xi},\sigma) \text{ for } g \in \Gamma,\end{aligned}\right\} \quad (5.31)$$

where $Ad_g\sigma$ is as specified in (4.29).

b) If $[p_1,p_2] \subseteq (0,p_0)$ is a second compact neighborhood of $\bar{p}$, and if f_2 is the function associated with it in the manner of part a), then f and f_2 coincide on the intersection of their domains.

c) If $\bar{p} \in (0,p_0)$, if $[p_1,p_2]$, $\mathcal{B}_\eta$, and f are as specified in part a), if $p \in (p_1,p_2)$, $\bar{\xi} \in B_\eta$, and if $\xi = p\xi_3 + \bar{\xi}$, then the critical points of $V_1(\xi,\cdot)$ lying in the open neighborhood $\rho(G_1 \times U_{2\varepsilon})$ of $G\gamma_0$ are in bijective correspondence with the critical points of $f(p,\bar{\xi},\cdot)$ in G_1.

Proof. Part a) follows from the symmetry properties of V_1, ρ, and ψ specified in Proposition 4.8, Lemma 5.11 and Proposition 5.17. Part b) follows from Proposition 5.17 b). By Lemma 5.14, $\rho(\sigma,\psi(p,\bar{\xi},\sigma))$ is a critical point for $V_1(\xi,\cdot)$ if and only if $(\sigma,\psi(p,\bar{\xi},\sigma))$ in $G_1 \times U_{2\varepsilon}$ is a critical point for $V_\rho(\xi,\cdot,\cdot)$. By Proposition 5.16, this is the case if and only if $(\sigma,\psi(p,\bar{\xi},\sigma))$ satisfy the two conditions stated therein. By Proposition 5.17 the pair satisfy the first condition. The second condition is the assertion that σ is a critical point for $f(p,\bar{\xi},\cdot)$. ∎

We obtain the following bifurcation problem over a finite-dimensional space.

5.19 Problem. *Let $\bar{p} \in (0,p_0)$. Let $[p_1,p_2]$, $\mathcal{B}_\eta$, and f be given by Theorem 5.18. For $p \in (p_1,p_2)$ and $\bar{\xi} \in \mathcal{B}_\eta$, determine the critical points for $f(p,\bar{\xi},\cdot)$.* ∎

Remarks.

1. The set of critical points for $f(p,0,\cdot)$ is G_1. So Problem 5.19 is a bifurcation problem in that you seek to determine how

the set of critical points for a function changes as you vary the parameter $\bar{\xi}$.

2. The bifurcation problem is finite-dimensional in that the critical points lie in G_1, a one-dimensional manifold with two components.

3. The reduction provided by Theorem 5.18 may be completed also in the case where the material comprising the rod is *not* hyperelastic (see also [2], p. 371). Moreover, a similar reduction for the special Cosserat rod is possible for an analogous set of problems.

4. By Theorem 5.18 b) the results of Problem 5.19 are independent of the choice of compact neighborhood about $\bar{p}$. However as p_2 approaches p_0, the diameter of $\mathcal{B}_\eta$ approaches zero, heralding the fact that p_0 is a singular point of the analysis.

In the subsequent sections we will examine only particular types of perturbations of ξ_0.

5.20 Problem. *Let*

$$S^1 = \{\ \xi \in \mathcal{J}\ |\ \|\xi\| = 1\ \},$$

where the norm is that inducted by (3.47). Let $\bar{\xi} \in \mathcal{K} \cap S^1$, *and let* $\bar{p} < p_0$. *For* $[p_1, p_2] \subseteq (0, p_0)$ *a compact neighborhood of* $\bar{p}$, *for* $p \in (p_1, p_2)$, *for* $\tau \in \mathbb{R}$ *such that* $\tau\bar{\xi} \in \mathcal{B}_\eta$, *and for* $\sigma \in G_1$, *define* $\ell(p, \tau, \sigma; \bar{\xi}) \in \mathbb{R}$ *by*

$$\ell(p, \tau, \sigma; \bar{\xi}) = f(p, \tau\bar{\xi}, \sigma), \tag{5.32}$$

where f *is given by (5.30). For* $\tau > 0$, *determine the critical points for* $\ell(p, \tau, \cdot\,; \bar{\xi})$. ∎

Problem 5.20 is tractable. It constitutes our mathematical formulation for the "pure orbit breaking" problem presented in the introduction. As [1], p. 324f and [2], p. 214f indicate, without restricting attention to special perturbations which preserve symmetry, Problem 5.19 remains an open question.

We conclude this section by indicating how Problem 4.25 reduces when $p = p_0$.

5.21 Proposition. *Let $p = p_0$. Let $T_{\gamma_0}\mathcal{M}$ and $\mathcal{F}$ be decomposed as in Theorem 5.10. Let $U_{2\varepsilon}$ and ρ be as specified as in Lemmas 5.11 and 5.15. Let*

$$W_{3\varepsilon} = W_3(\gamma_0) \cap U_{2\varepsilon} \text{ and } W_{4\varepsilon} = W_4(\gamma_0) \cap U_{2\varepsilon}.$$

For $\xi \in \mathcal{J}$, $\sigma \in G_1$, $Z_3 \in W_{3\varepsilon}$. and $Z_4 \in W_{4\varepsilon}$, define

$$V(\xi,\sigma,Z_3,Z_4) = V_1(\xi,\rho(\sigma,\Omega(Z_3+Z_4))). \tag{5.33}$$

a) Then the critical points of $V_1(\xi,\cdot)$ in the neighborhood $\mathcal{U}(G\gamma_0)$ are in bijective correspondence with those of $V(\xi,\cdot,\cdot,\cdot)$ in $G_1 \times W_{3\varepsilon} \times W_{4\varepsilon}$.

b) The point (σ,Z_3,Z_4) is a critical point for $V(\xi,\cdot,\cdot,\cdot)$ if and only if, for $Z = Z_3+Z_4$ and $Y_4 \in W_4$,

$$(1)\quad K(\Psi(\xi,\rho(\sigma,\Omega(Z)),M_{\Omega(Z)}Y_4) = 0, \tag{5.34}$$

(2) (σ,Z_3) is a critical point for the function defined on G_1 X $W_{3\varepsilon}$ given by

$$h(\sigma,Z) = V(\xi,\sigma,Z,Z_4).$$

Proof. Lemma 5.11 continues to hold when we decompose $T_{\gamma_0}\mathcal{M}$ as specified in (5.17). Consequently, Lemma 5.14 continues to hold when $U_{2\varepsilon}$ is decomposed as $U_{2\varepsilon} = W_{3\varepsilon} \oplus W_{4\varepsilon}$, a Γ-invariant decomposition. Using these lemmas, the proof of the proposition then follows in the manner of the proof of Proposition 5.16. ∎

5.22 Proposition. *Let $p = p_0$. Assume the hypotheses of Theorem 5.10 hold. Let $U_{2\varepsilon} = W_{3\varepsilon} \oplus W_{4\varepsilon}$ as in Proposition 5.21. Then* *there is an interval* $\mathcal{I}_\alpha$ *about 0 in* $\mathbb{R}$*, a neighborhood* $\mathcal{B}_\eta$ *of 0 in* $\mathcal{K}$ *and a smooth function* ψ *defined on* $\mathcal{I}_\alpha \times \mathcal{B}_\eta \times G_1 \times W_{3\varepsilon}$ *and taking values in* $W_{4\varepsilon}$ *such that, if* $\lambda \in \mathcal{I}_\alpha$, $\bar{\xi} \in B_\eta$, *and* $(\sigma, Z_3, Z_4) \in G_1 \times W_{3\varepsilon} \times W_{4\varepsilon}$, *(5.34) holds if and only if*

$$Z_4 = \psi(\lambda, \bar{\xi}, \sigma, Z_3).$$

Moreover, ψ *satisfies the additional properties:*

$$\left.\begin{array}{ll} a) & \psi(\lambda, 0, \sigma, 0) = 0 \quad \text{for } \sigma \in G_1, \\ b) & \psi(\lambda, \sigma_1\bar{\xi}, \sigma_1\sigma, Z_3) = \psi(\lambda, \bar{\xi}, \sigma, Z_3) \quad \text{for } \sigma_1 \in G_1, \\ c) & \psi(\lambda, g\bar{\xi}, Ad_g\sigma, \mathcal{B}(g)Z_3) = \mathcal{B}(g)\psi(\lambda, \bar{\xi}, \sigma, Z_3) \quad \text{for } g \in \hat{\Gamma} \end{array}\right\} \quad (5.35)$$

where $\mathcal{B}(g)$ *and* $Ad_g\sigma$ *are given by (2.23) and (4.29), respectively.*

Proof. The proposition follows from Theorem 5.10, Proposition 5.21, and the Implicit Function Theorem, in the manner of Proposition 5.17. ∎

Problem 4.25 then reduces for the case $p = p_0$ (see [4], p. 331).

5.23 Lemma. *Let* f_0 *and* $W_3(\gamma_0)$ *be given by Theorem 5.10. Let* B *be given by (4.16).*

a) For $\underline{X} = (X_1, X_2) \in \mathbb{R}^2$*, define* $\mathbf{l}(\underline{X}) \in W_3(\gamma_0)$ *by*

$$\mathbf{l}(\underline{X}) = X_\alpha \hat{\underline{e}}_\alpha f_0 \, .$$

Then $\mathbf{l}$ *is a linear isomorphism of* $\mathbb{R}^2$ *onto* $W_3(\gamma_0)$ *for which*

$$B(\mathbf{l}(\underline{X}), \mathbf{l}(\underline{X})) = \|\underline{X}\|_{\mathbb{R}^2} \, ,$$

b) Let $Q\underline{X}$ *denote the usual action of SO(3) on* R^3*. Then* $\mathbb{R}^2$ *is an invariant subspace for* Γ*. If* $\underline{X} \in \mathbb{R}^2$*, if* $Q \in \Gamma$*, if* $g = (Q,Q) \in \hat{\Gamma}$*, and if* $\mathcal{B}(g)$ *is given by (2.23), then*

$$\mathbf{l}(Q\underline{X}) = \mathcal{B}(g)\mathbf{l}(\underline{X}).$$

Proof. The lemma follows from (4.16), (4.27), and the definitions of f_0 and $W_3(\gamma_0)$ given in Theorem 5.10. ∎

5.24 Theorem. *Let* $p = p_0$. *Let* $\mathcal{I}_\alpha$, $\mathcal{B}_{\eta'}$, $W_{3\varepsilon}$, ρ, *and* ψ *be as specified in Proposition 5.22. Let* $U_\varepsilon \subseteq \mathbb{R}^2$ *be the* ε*-ball about 0 in* $\mathbb{R}^2$ *associated with* $W_{3\varepsilon}$ *by Lemma 5.23. For* $(\lambda,\bar{\xi},\sigma,\underline{X}) \in \mathcal{I}_\alpha \times \mathcal{B}_\eta \times G_1 \times U_\varepsilon$, *define*

$$f_2(\lambda,\bar{\xi},\sigma,\underline{X}) = V(\xi,\sigma,1(\underline{X}),\psi(\lambda,\bar{\xi},\sigma,1(\underline{X}))), \tag{5.36}$$

where $\xi = (p_0+\lambda)\xi_3 + \bar{\xi}$, V *is as specified in (5.33) and* 1 *is given by Lemma 5.23.*

a) Then the critical points of $V_1(\xi,\cdot)$ *in* $\mathcal{U}_1(G\gamma_0)$ *are in bijective correspondence with the critical points of* $f_2(\lambda,\bar{\xi},\cdot,\cdot)$ *in* $G_1 \times U_\varepsilon$.

b) The function f_2 *satisfies*

$$\left.\begin{array}{ll} (1) & f_2(\lambda,o,\sigma,\underline{X}) = f_2(\lambda,0,1,\underline{X}), \\ (2) & f_2(\lambda,\sigma_1\bar{\xi},\sigma_1\sigma,\underline{X}) = f_2(\lambda,\bar{\xi},\sigma,\underline{X}), \quad \sigma_1 \in G_1 \\ (3) & f_2(\lambda,g\bar{\xi},Ad_g\sigma,Q\underline{X}) = f_2(\lambda,\bar{\xi},\sigma,\underline{X}), \quad g \in \Gamma, \end{array}\right\} \tag{5.37}$$

where $g = (Q,Q)$, *and* $Ad_g\sigma$ *is given by (4.29).*

Proof. The theorem follows from Propositions 5.21 and 5.22 and Lemma 5.23 in the manner of the proof of Theorem 5.18. ∎

We obtain the following finite-dimensional bifurcation problem.

5.25 Problem. *Let* $\mathcal{I}_\alpha$ *and* $\mathcal{B}_\eta$ *be given by Theorem 5.24. Let* $\lambda \in \mathcal{I}_\alpha$, $\bar{\xi} \in \mathcal{B}_{\eta'}$, *and let* $\xi = (p_0+\lambda)\xi_3 + \bar{\xi}$. *Determine the set of critical points for* $f_2(\lambda,\bar{\xi},\cdot,\cdot)$. ∎

Remarks analogous to those made after Problem 5.19 hold also for this problem. Here, the set of critical points for $f_2(0,0,\cdot,\cdot)$ is $G_1 \times U_\varepsilon$, a three-dimensional manifold.

Analogous to Problem 5.20 we have the following special case.

5.26 Problem. *Let* $\bar{\xi} \in \mathcal{K} \cap S^1$ *as in Problem 5.20. Let* λ, $\tau > 0$ *be sufficiently small that* $\lambda \in \mathcal{I}_\alpha$ and $\tau\bar{\xi} \in \mathcal{B}_\eta$. *Define* $\ell_2(\lambda,\tau,\sigma,\underline{X};\bar{\xi}) \in \mathbb{R}$ *by*

$$\ell_2(\lambda,\tau,\sigma,\underline{X};\bar{\xi}) = f_2(\lambda,\tau\bar{\xi},\sigma,\underline{X}), \tag{5.38}$$

where f_2 *is given by (5.36). Determine the critical points for* $\ell_2(\lambda,\tau,\cdot,\cdot;\bar{\xi})$ *in* $G_1 \times U_\varepsilon$. ∎

Problem 5.26 constitutes a tractable form of the "full" rod problem which we presented in the introduction. For the case $\tau = 0$, Problem 5.26 specializes to a formulation of the "symmetric buckling" problem, which we also discussed in the introduction. We present it here for completeness.

5.27 Problem. *Let* f_2 *be given by (5.36). For* λ *sufficiently small, determine the critical points for* $f_2(\lambda,0,1,\cdot)$ *in* U_ε *(see Lemma 6.3).* ∎

Problem 5.27 is analogous to the one considered in [4], with the material symmetry group being O(2), as opposed to a dihedral group.

VI. THE ANALYSIS OF THE REDUCED PROBLEM

In this section we analyze Problem 5.20. First, we indicate how the perturbing load $\bar{\xi}$ determines how we analyze the problem. We then classify the types of perturbing loads arising in Problem 5.20 and illustrate the classification.

While we examine Problem 5.20 in detail, we consider only the particular case of Problem 5.26 presented as Problem 5.27, where the perturbing load $\bar{\xi} = 0$. This case represents the symmetrically buckled problem discussed in the introduction. It was studied in [4] and [9]. We will indicate how the analysis presented in this work relates to these works. In Section 8, we will comment on what happens when $\bar{\xi} \neq 0$ at $p = p_0$. We reserve specific details for a future presentation.

VI.1. *The Symmetrically Perturbed Problem*

As indicated in [3], p. 220, how we proceed to solve Problems 5.20 and 5.26 depends significantly upon whether the perturbing load $\bar{\xi}$ maintains or breaks the symmetry of the trivial load ξ_0. To formalize this idea, recall from Lemma 4.22 that G is the group of elements of Π which leave ξ_3 invariant under the action given by (3.39).

6.1 Definition. *Let* $\xi_0 = p\xi_3$, *and* $\eta \in \mathcal{J}$. *Say* η **maintains** *the symmetry of* ξ_0 *if*

$$\{g \in G \mid g\eta = \eta\} = G.$$

Otherwise, say η **breaks** *the symmetry of* ξ_0. *If* η *maintains the symmetry of* ξ_0, *call* η *a* **symmetric perturbation** *of* ξ_0. ∎

For the constrained problem η maintains symmetry when $\eta = \lambda\xi_3$. This case produces some of the symmetrically perturbed bifurcation problems which were examined in [9] (see p. 298, case B). They are the counterparts to the problems considered in [4].

If we analyze Problems 5.20 and 5.26 subject to a symmetric perturbation $\eta = \lambda\xi_3$ using the symmetric bifurcation theory, we do not improve upon the results obtained in [9]. However, it is instructive to summarize this method of analysis, in order to see how it relates to the more established techniques.

6.2 Lemma. *If* $p < p_0$ *then the orbit* $G\gamma_0$ *in* $\mathcal{M}_1$ *is the set of equilibrating configurations for Problem 4.25 for* $\xi = p\xi_3$ *in the neighborhood* $U(G\gamma_0)$.

Proof. Since $f(p,0,\cdot)$ is constant on G_1, $G_1 = \mathrm{crit} f(p,0,\cdot)$. So the lemma follows from Theorem 5.18, (5.29), and Lemma 5.11. ∎

So when $p < p_0$, a symmetric bifurcation produces no bifurcation from the trivial orbit $G\gamma_0$, in the sense that there is a one-to-one correspondence between the orbit $G\gamma_0$ and the set of equilibrating configurations for the perturbed problem $(p+\lambda)\xi_3$ which lie nearby $G\gamma_0$.

When $p = p_0$ and $\bar{\xi} = 0$, we reduce to the problem which was considered in [9], and in [4] for the case of prismatic rods. As these results are subsumed by those of [9], and as the techniques are those used in [4], we present them summarily, so that the reader can relate the works.

6.3 Proposition. *Let* $p = p_0$ *and* $\bar{\xi} = 0$. *Let* f_2 *be given by (5.36).*

a) Then $f_2(\lambda,0,\sigma,\underline{X}) = f_2(\lambda,0,1,\underline{X}) \equiv n(\lambda,\underline{X})$. *Moreover,*

$$n(\lambda,Q\underline{X}) = n(\lambda,\underline{X}),$$

where $Q \in \Gamma$ *and acts on* $\mathbb{R}^2$ *in a manner isomorphic to* $O(2)$ *with its usual action.*

b) Define $\underline{G}(\lambda,\underline{X}) \in \mathbb{R}^2$ *by, for* $\underline{w} \in \mathbb{R}^2$,

$$\underline{G}(\lambda,\underline{X})\cdot\underline{w} = [T_2 n(\lambda,\underline{X})]\underline{w}\ .$$

(1) Then under the action of Γ the orbits of the zeros of $\underline{G}(\lambda,\underline{X})$ are in bijective correspondence with the zeros of

$$\underline{N}(\lambda,\underline{X}) = (-\lambda + \delta v)\underline{X} \in \mathbb{R}^2,$$

where $v = \|\underline{X}\|^2$ and $\delta \neq 0$ is determined by Hypothesis 5.7.

(2) As λ passes through zero, $\underline{N}$ undergoes (an O(2)-symmetric) pitchfork bifurcation.

Proof. Part a) follows from (5.37) and the action of Γ on $\mathbb{R}^2$ specified in Lemma 5.23. Using the chain rule, (5.36), and (5.26) in the manner of the proof of Proposition 5.16 we obtain

$$\underline{G}(\lambda,\underline{X}) = K\big(\Psi((p_0+\lambda)\xi_3,\rho(1,Z(\lambda,\underline{X}))),M_{Z(\lambda,\underline{X})}(\partial(Z(\lambda,\underline{X})/\partial x_\alpha))\underline{e}_\alpha ,$$

where $Z(\lambda,\underline{X}) = 1(\underline{X}) + \psi(\lambda,0,1,1(\underline{X})) \in U_2(\gamma_0)$, given by Proposition 5.22 and Lemma 5.23. By a computation based upon (5.35), Lemma 5.11, Lemma 4.16, (4.14), and the relation between the action of Γ and O(2) on $\mathbb{R}^2$ described in Lemma 5.23, $\underline{G}(\lambda,Q\underline{X}) = Q\underline{G}(\lambda,\underline{X})$ for $Q \in O(2)$. By [30], it follows that

$$\underline{G}(\lambda,\underline{X}) = g(\lambda,v)\underline{X} ,$$

for $g(\lambda,v)$ a real-valued function, and $v = \|\underline{X}\|^2$. A computation using Lemma 5.5 gives (see also [4])

$$g(0,0) = \int_I \left[2H_{\tau_1}(S,0)f_0'f_0' - p_0f_0f_0\right]dS = 0 ,$$

where the last equality arises from (5.16). Further,

$$g_\lambda(0,0) = -\int_I f_0^2 dS = -1 ,$$

because f_0 is a normalized eigenfunction. Finally, using Lemma 5.5 we obtain

$$\mathcal{q}_V(0,0) = \int_I \left[2H_{\tau_1\tau_1}(S,0)(f_0')^4+(p_0/6)f_0^4\right]dS > 0 ,$$

where the last inequality arises from Hypothesis 5.7. If $c = \mathcal{q}_V(0,0)$, taking $\delta = \operatorname{sgn}(c/|c|)$ gives

$$\underline{G}(\lambda,\underline{X}) = \underline{N}(\lambda,\underline{X}) + \text{HOT} .$$

From the O(2)-symmetric theory developed in [30], it follows that $\underline{G}$ is O(2)-equivalent to $\underline{N}$, and that their zero sets are in bijective correspondence, modulo O(2). The analysis of the normal form $\underline{N}(\lambda,\underline{X})$ (see [30]) gives the pitchfork. ∎

The procedure used to obtain Proposition 6.3 was used in [4] to study prismatic rods. In that work, the symmetry group Γ was a dihedral group. In this work, Γ is isomorphic to O(2). As we indicated before, the results obtained are subsumed by those obtained in [9] using the bifurcation theory of Crandall and Rabinowitz.

Original results do arise, however, when we consider problems where the perturbing load breaks the symmetry of ξ_3. We examine this situation in detail.

VI.2. *The Critical Manifolds for the Symmetry-Breaking Loads*

Consider Problem 5.20, where $\bar{\xi} \in \mathcal{K} \cap S^1$. To analyze this problem, we expand $\ell(p,\tau,\sigma;\bar{\xi})$ of (5.32) to third order in τ. We accomplish the expansion using the following lemma.

6.4 Lemma. *Let* $\bar{p} < p_0$ *and* $\bar{\xi} \in \mathcal{K} \cap S^1$. *Let*

$$\psi(p,\tau,\sigma;\bar{\xi}) \equiv \psi(p,\tau\bar{\xi},\sigma) \tag{6.1}$$

be given by Proposition 5.17. Set

$$u_1(p,\sigma;\bar\xi) = \psi_\tau(p,\tau,\sigma;\bar\xi)_{|\tau=0} , \tag{6.2}$$

where the subscript τ denotes a partial derivative. Let L_p be given by (5.5). Then for $W_2 \in U_2(\gamma_0)$,

$$K(L_p u_1(p,\sigma;\bar\xi),W_2) = K(\beta(\sigma^{-1}\bar\xi,\gamma_0),W_2) , \tag{6.3}$$

or equivalently,

$$L_p u_1(p,\sigma;\bar\xi) = \Pi_{2e}\beta(\sigma^{-1}\bar\xi,\gamma_0) , \tag{6.4}$$

(independent of p), where Π_{2e} is the projection of $\mathcal{F}$ onto $\mathcal{F}_{2e}(\gamma_0)$ given by (5.4).

Proof. For $\bar p < p_0$, and p and τ sufficiently small, Proposition 5.17 gives, for $\xi = p\xi_3 + \tau\bar\xi$,

$$K\big(\Psi(\sigma^{-1}\xi,\rho(1,\Omega(\psi(p,\tau,\sigma;\bar\xi))),M_{\Omega(\psi(p,\tau,\sigma;\bar\xi))}T\Omega(\psi(p,\tau,\sigma;\bar\xi))W_2\big) \equiv 0 ,$$

for $W_2 \in U_2(\gamma_0)$ arbitrary. Differentiating with respect to τ and evaluating at $\tau = 0$ using (3.35), (4.12), Lemma 4.23, (5.5), Lemma 5.12, (5.26), and (5.29) gives

$$K(L_p u_1(p,\sigma;\bar\xi) - \beta(\sigma^{-1}\bar\xi,\gamma_0),W_2) = 0.$$

Equation (6.4) follows from the nondegeneracy of K on $U_2(\gamma_0) \times \mathcal{F}_{2e}(\gamma_0)$, $u_1(p,\sigma;\bar\xi) \in U_2(\gamma_0)$, and Theorem 5.9 b). ∎

We may now begin to characterize the critical points of $\ell(p,\tau,\sigma;\bar\xi)$ on G_1.

6.5 Proposition. *Let $\bar p < p_0$ and $\bar\xi \in \mathcal{K} \cap S^1$. Let $\ell(p,\tau,\sigma;\bar\xi)$ be given by (5.32) on some compact neighborhood of $\{(\bar p,0)\} \times G_1$.*

a)

$$\ell(p,\tau,\sigma;\bar\xi) = f(p,0,1) + \tau h(p,\tau,\sigma;\bar\xi) , \tag{6.5}$$

and for $\tau \neq 0$, $\sigma \in \operatorname{crit}\ell(p,\tau,\cdot;\bar\xi)$ if and only if $\sigma \in \operatorname{crit} h(p,\tau,\cdot;\bar\xi)$.

b) *For*

$$A_{\bar{\xi}}(\sigma) = trk(\sigma^{-1}\bar{\xi}, \gamma_0), \tag{6.6}$$

$$h(p,\tau,\sigma) = -A_{\bar{\xi}}(\sigma) - \frac{\tau}{2} K(L_p u_1(\lambda,\sigma;\bar{\xi}), u_1(\lambda,\sigma;\bar{\xi})) + O(\tau^2). \tag{6.7}$$

Proof.

a) Equation (6.5) follows from (5.31), (5.32) and the Mean Value Theorem. The bijective correspondence of the sets of critical points then follows from (6.5).

b) Taylor's theorem and the compactness of G_1 give

$$h(p,\tau,\sigma;\bar{\xi}) = h(p,0,\sigma;\bar{\xi}) + \tau h_\tau(p,0,\sigma;\bar{\xi}) + O(\tau^2), \tag{6.8}$$

where the $O(\tau^2)$ is bounded on compact neighborhoods of $\{(\bar{p},0)\} \times G_1$. By (6.5),

$$\left.\begin{aligned} h(p,0,\sigma;\bar{\xi}) &= \ell_\tau(p,0,\sigma;\bar{\xi}) \\ h_\tau(p,\tau,\sigma;\bar{\xi}) &= \tfrac{1}{2}\ell_{\tau\tau}(p,0,\sigma;\bar{\xi}) \end{aligned}\right\} \tag{6.9}$$

By (5.30) through (5.32), (6.1) and (6.2), for $\xi = p\xi_3 + \tau\bar{\xi}$,

$$\left.\begin{aligned} \ell_\tau(p,\tau,\sigma;\bar{\xi}) = {} & [T_1V_1(\sigma^{-1}\xi, \rho(1,\Omega(\psi(p,\tau,\sigma;\bar{\xi}))]\sigma^{-1}\xi + \\ & [T_2V_1(\sigma^{-1}\xi,\rho(1,\Omega(\psi))]T_2\rho(1,\Omega(\psi)T\Omega(\psi)\psi_\tau \end{aligned}\right\}. \tag{6.10}$$

By (4.1), (4.5), (4.11), and (5.19), (6.10) yields

$$\left.\begin{aligned} \ell_\tau(p,\tau,\sigma;\bar{\xi}) = {} & -\operatorname{trk}(\sigma^{-1}\bar{\xi},\rho(1,\Omega(\psi(p,\tau,\sigma;\bar{\xi})))) + \\ & K\big(\Psi(\sigma^{-1}\xi,\rho(1,\Omega(\psi))), M_{\Omega(\psi)}T\Omega(\psi)\psi\big)_\tau. \end{aligned}\right\} \tag{6.11}$$

Since $\psi(p,\tau,\sigma;\bar{\xi}) \in U_2(\gamma_0)$, Proposition 5.17 implies that the second term of (6.11) is zero. Hence,

$$\ell_\tau(p,\tau,\sigma;\bar{\xi}) = -\operatorname{trk}(\sigma^{-1}\bar{\xi},\rho(1,\Omega(\psi(p,\tau,\sigma;\bar{\xi})))). \tag{6.12}$$

By (4.1), (4.3), and (5.19),

$$\ell_{\tau\tau}(p,\tau,\sigma;\bar{\xi}) = -K(\beta(\sigma^{-1}\bar{\xi},\rho(1,\Omega(\psi))),M_{\Omega(\psi)}T\Omega(\psi)\psi_\tau) \quad . \tag{6.13}$$

Evaluating (6.12) and (6.13) at $\tau = 0$ using Lemma 5.11, (5.25), (5.29), (6.2), and (6.6) gives

$$\left.\begin{aligned} \ell_\tau(p,\tau,\sigma;\bar{\xi}) &= -A_{\bar{\xi}}(\sigma) \\ \ell_{\tau\tau}(p,\tau,\sigma;\bar{\xi}) &= -K\big(L_p u_1(p,\sigma;\bar{\xi}),u_1(p,\sigma;\bar{\xi})\big) \end{aligned}\right\}. \tag{6.14}$$

Equations (6.8), (6.9), and (6.14) then give (6.7). ∎

The expansion of h given by (6.7) indicates the importance of the set of critical points for $A_{\bar{\xi}}(\sigma)$ in resolving Problem 5.20.

6.6 Notation. *Let $\mathbb{K}$ denote either the space $\mathbb{R}$, $\mathcal{C}$, or $\mathcal{F}$. Let X be an arbitrary space. Let $F(\sigma,x) \in \mathbb{K}$ be a smooth function on $G_1 \times X$. Denote by $F_\sigma(\sigma,x)$ the element in $\mathbb{K}$ given by*

$$F_\sigma(\sigma,x) = [T(F(\cdot,x))(\sigma)]TL_\sigma(1)\hat{\underline{e}}_3 \ ,$$

where $\hat{\underline{e}}_3 \in T_1G_1$ is the generator of the Lie algebra for G_1 and L_σ is left translation. ∎

6.7 Lemma. If $\mathcal{I}(\sigma) = \sigma^{-1}$, then

$$\frac{\partial}{\partial\sigma}\mathcal{I}(\sigma) = -\sigma^{-1}\sigma\hat{\underline{e}}_3\sigma^{-1} = -\hat{\underline{e}}_3\sigma^{-1} \ . \tag{6.15}$$

Proof. See [4], Chapter 4. ∎

6.8 Definition. *Let $\bar{\xi} \in \mathcal{J}$. Denote by $S_{\bar{\xi}}$ the set of critical points for $A_{\bar{\xi}}$ in G_1. Call $S_{\bar{\xi}}$ the critical manifold for $\bar{\xi}$.* ∎

The critical points we seek lie in a neighborhood of $S_{\bar{\xi}}$.

6.9 Lemma. *Let $h(p,\tau,\sigma;\bar{\xi})$ be given by (6.7). Let $\mathcal{U}(S_{\bar{\xi}})$ be a neighborhood of $S_{\bar{\xi}}$ in G_1. Then there is a $T_0 > 0$ such that for*

$0 < \tau < T_0$, the critical points for $h(p,\tau,\cdot;\bar{\xi})$ lie in $\mathcal{U}(S_{\bar{\xi}})$.

Proof. From Proposition 6.5,

$$h_{\sigma}(p,\tau,\sigma;\bar{\xi}) = -\left(A_{\bar{\xi}}(\sigma)\right)_{\sigma} + O(\tau) , \qquad (6.16)$$

where the $O(\tau)$ approximation is bounded on compact neighborhoods of $\{(\bar{p},o)\} \times G_1$. Thus for some $c_1 > 0$,

$$|h_{\sigma}(p,\tau,\sigma;\bar{\xi})| \geq \left|\left(A_{\bar{\xi}}(\sigma)\right)_{\sigma}\right| - \tau c_1 . \qquad (6.17)$$

Since $G_1 \setminus \mathcal{U}(S_{\bar{\xi}})$ is compact and disjoint from $S_{\bar{\xi}}$, there is a $c_2 > 0$ which bounds from below the first term of the right hand side of (6.17) for all $\sigma \in G_1 \setminus \mathcal{U}(S_{\bar{\xi}})$. Taking $0 < T_0 < (c_2/c_1)$ gives the lemma. ■

6.10 Lemma. *If $\sigma \in G_1$, then $\sigma \in S_{\bar{\xi}}$ if and only if*

$$k(\sigma^{-1}\bar{\xi},\gamma_0):\hat{\underline{e}}_3 = 0, \qquad (6.18)$$

or equivalently,

$$k(\sigma^{-1}\bar{\xi},\gamma_0) \in sym(E^3) \oplus o(2)^{\perp}, \qquad (6.19)$$

where $sym(E^3)$ is the space of symmetric linear operators on E^3, and $o(2)^{\perp} = span\{\hat{\underline{e}}_1, \hat{\underline{e}}_2\}$.

Proof. From the antisymmetry of $\hat{\underline{e}}_3$ and the definition of k,

$$\left.\begin{aligned}\left(A_{\bar{\xi}}(\sigma)\right)_{\sigma} &= [T_1(\mathrm{trk})(\sigma^{-1}\bar{\xi},\gamma_0)]\hat{\underline{e}}_3 = -\mathrm{trk}(\hat{\underline{e}}_3\sigma^{-1}\bar{\xi},\gamma_0)\\ &= \mathrm{trk}(\sigma^{-1}\bar{\xi},\hat{\underline{e}}_3\gamma_0) = k(\sigma^{-1}\bar{\xi},\gamma_0):\hat{\underline{e}}_3.\end{aligned}\right\} \qquad (6.20)$$

So σ is a critical point for $A_{\bar{\xi}}$ if and only if (6.18) holds. ■

$S_{\bar{\xi}}$ has a physical interpretation. In light of Definition 3.15 and the remark after Theorem 4.20, Lemma 6.10 asserts that $S_{\bar{\xi}}$

consists of those elements of G_1 which "rotate" $\bar{\xi}$ into a load whose total torque of the rod in the configuration γ_0 can be balanced by the forces arising from the constraint that the ends of the rod lie on the $\underline{e}_3$ axis.

VI.3. *The Classification of the Critical Manifolds*

Lemma 6.10 will allow us to establish a system for classifying the perturbing loads which we can use to solve Problem 5.20. To obtain the classification we first transform the perturbing load into a "canonical form". To define the form and to establish the relation between a load in $\mathcal{K}$ and its canonical form we exploit the work of [31].

6.11 Notation.

a) If $T \in L(E^3)$, let T_{ij} denote the i,j^{th} component of the matrix representing T with respect to the basis $\underline{e}_j$, $j = 1, 2, 3$. In particular denote

$$T_{11} = A, \quad T_{12} = B, \quad T_{21} = C, \quad T_{22} = D.$$

b) If $Q \in SO(2)$, write $Q = R_\phi$, where ϕ is the angle of counterclockwise rotation about the $\underline{e}_3$ axis Q produces by its usual action on E^3. Set

$$Q_{11} = x = \cos\phi, \text{ and } Q_{21} = y = \sin\phi.$$

c) Let G act on $L(E^3)$ as specified in Definition 2.4. Denote by $G\cdot T$ the orbit of T in $L(E^3)$:

$$G\cdot T = \{\ S \in L(E^3) \mid S = g\cdot T \text{ for some } g \in G\ \}. \qquad (6.21)$$

Similarly, for G_1 and Γ given by (4.28), denote by $G_1\cdot T$ and $\Gamma\cdot T$ the orbits generated by these subgroups of G. ∎

6.12 Lemma. *Let $\alpha = 1, 2$ and $j = 1, 2, 3$. In $L(E^3)$ set*

$$V_1 = \{ T \mid T_{j3} = T_{3j} = 0 \} \approx L(2) ,$$
$$V_2 = \{ T \mid T_{j\alpha} = T_{33} = 0 \} \approx \mathbb{R}^2 ,$$
$$V_3 = \{ T \mid T_{\alpha j} = T_{33} = 0 \} \approx \mathbb{R}^2 ,$$
$$V_4 = \{ T \mid T_{j\alpha} = T_{\alpha j} = 0 \} \approx \mathbb{R} .$$

Then $L(E^3)$ decomposes as a vector space into the direct sum

$$L(E^3) = V_1 \oplus V_2 \oplus V_3 \oplus V_4 .$$

a) The decomposition is invariant under G.

b) If $\Pi_j : L(E^3) \longrightarrow V_j$ denotes a projection in the direct sum, then $g\cdot(\Pi_j T) = \Pi_j(g\cdot T)$.

c) G *acts irreducibly on each factor: for each factor there is no proper, nontrivial subspace invariant under* G.

Proof. The factors are subspaces of $L(E^3)$ with trivial intersection, so the sum is direct, and forms a subspace. Representing $T \in L(E^3)$ as a matrix relative to $\{\underline{e}_j\}$ we may write

$$\begin{pmatrix} a & b & r \\ c & d & s \\ g & h & f \end{pmatrix} = \begin{pmatrix} a & b & 0 \\ c & d & 0 \\ 0 & 0 & 0 \end{pmatrix} + \begin{pmatrix} 0 & 0 & r \\ 0 & 0 & s \\ 0 & 0 & 0 \end{pmatrix} + \begin{pmatrix} 0 & 0 & 0 \\ 0 & 0 & 0 \\ g & h & 0 \end{pmatrix} + \begin{pmatrix} 0 & 0 & 0 \\ 0 & 0 & 0 \\ 0 & 0 & f \end{pmatrix} .$$

Hence, $L(E^3)$ is the direct sum asserted. A computation based upon Lemma 4.22 gives the three properties. ∎

We develop the canonical forms for $\bar{\xi} \in \mathcal{K}$ using some results from the theory of group representations. We examine two orbits for $\Pi_1 T$: $G\cdot\Pi_1 T$ and $SG\cdot\Pi_1 T$, where SG is the component of G which contains the identity. We call them the unoriented and oriented orbits for $\Pi_1 T$, respectively. Their significance will be seen in Proposition 6.17.

By (4.28) and Lemma 4.28, $G = G_1\Gamma$ and $SG = SG_1 S\Gamma$. We examine the contributions of Γ and $S\Gamma$ to the orbits of $\Pi_1 T$ in the next two lemmas.

6.13 Lemma. *Let $L(2)$ be the space of 2×2 matrices. Let $O(2)$ act on $L(2)$ by adjoint action, $Ad_Q L = QLQ^T$. Let $Sym(2)$ denote the ($O(2)$ invariant) subspace of symmetric matrices. Define*

$$\begin{aligned} \Phi : Sym(2) &\longrightarrow \mathbb{C} \times \mathbb{R} \\ S &\longmapsto \phi(S) = (z,a) \ , \end{aligned}$$

where $z = r + is$, $r = (A-D)/2$, $s = B$, $a = (A+D)/2$, and A, B, D are given by Notation 6.11. Let $O(2)$ act on $\mathbb{C} \times \mathbb{R}$ by

$$Q(z,a) = \left\{ \begin{array}{ll} (e^{i2\phi}z,a) & \text{if } Q = R_\phi \in SO(2) \\ (\bar{z},a) & \text{if } Q = J \end{array} \right\} , \qquad (6.22)$$

where R_ϕ represents the standard action of $SO(2)$ on $\mathbb{R}^2$. Then

a) *Φ is a linear isomorphism, and $\Phi(Ad_Q S) = Q(\Phi(S))$.*

b) *$trS = 2a$, and $|z|^2 = a^2 - detS$.*

Proof. Part a) follows from a computation on Sym(2) (see [34] §2). Part b) follows from the definitions of the quantities involved. ∎

6.14 Lemma. *Let $S\Gamma$ be the component of Γ containing the identity. Let $S \in V_1$ and $S^T = S$. Suppose $trS \equiv 2a$. Then there is a unique element $N \in V_1$ for which*

$$\left. \begin{array}{l} 1) \ N \text{ is diagonal} \\ 2) \ \Gamma\cdot N = \Gamma\cdot S = S\Gamma\cdot S = S\Gamma\cdot N \\ 3) \ trN = 2a \\ 4) \ tr(JN) \equiv -2d, \text{ where } d \geq 0 \text{ and satisfies} \\ \qquad a^2 - d^2 = det_2 S, \end{array} \right\} \qquad (6.23)$$

where $det_2 S$ is the determinant of S, viewed as an element of $Sym(2)$ under the isomorphism of Lemma 6.12. In particular, if $det_2 S \geq 0$, and $trS \geq 0$, $0 \leq d \leq a$.

Proof. Let Φ be given by Lemma 6.13, and set $\Phi(S) = (z,a)$. Then $\Phi(\Gamma\cdot S) = O(2)\cdot(z,a)$, and $\Phi(S\Gamma\cdot S) = SO(2)\cdot(z,a)$. Since $(\bar{z},a) \in SO(2)\cdot(z,a)$ by (6.22), $\Gamma\cdot S = S\Gamma\cdot S$. Set $d = |z|$, so d is real and not negative. Set $N = \Phi^{-1}(d,a)$. Then N satisfies (6.23). Moreover, if N_2 satisfies properties 1) and 2) of (6.23), then a computation shows that either $N_2 = N$, or the elements of N_2 are a permutation of the diagonal elements of N. Property 3) then uniquely specifies N. Finally, Lemma 6.13 implies that $d^2 + \det_2 S = a^2$. So if $\det_2 S \geq 0$, $a \geq d \geq 0$. ∎

We now examine the contributions of G_1 and SG_1 to the orbits of $\Pi_1 T$. If $T \in L(E^3)$, if $Q \in SO(2)$, and if $g = (Q,1) \in G_1$, a computation gives

$$\sigma\cdot T:\hat{\underline{e}}_3 = -x(B - C) + y(A + D), \tag{6.24}$$

$$\mathrm{tr}_2(\sigma\cdot\Pi_1 T) = x(A + D) + y(B - C), \tag{6.25}$$

$$J\sigma\cdot T:\hat{\underline{e}}_3 = x(B + C) + y(A - D), \tag{6.26}$$

$$\mathrm{tr}_2(J\sigma\cdot\Pi_1 T) = -x(A - D) + y(B + C). \tag{6.27}$$

The equations produce the following lemma.

6.15 Lemma. *Let $T \in L(E^3)$. Set*

$$\left.\begin{aligned} M_1(T) &= \sqrt{(A + D)^2 + (B - C)^2}, \\ M_2(T) = M_1(JT) &= \sqrt{(A - D)^2 + (B + C)^2}. \end{aligned}\right\} \tag{6.28}$$

Let $\det_2(\Pi_1 T)$ denote the determinant of $\Pi_1 T \in V_1$, viewed under the isomorphism between V_1 and $L(2)$.

a) (oriented orbits) Let SG_1 be the component of G_1 containing the identity. Then there is an element $S \in V_1$ such that

$$\left.\begin{array}{ll} 1) & S \text{ is symmetric} \\ 2) & trS = M_1(T) \geq 0 \\ 3) & det_2 S = det_2(\Pi_1 T) \\ 4) & SG_1 \cdot S = SG_1 \cdot (\Pi_1 T) \end{array}\right\} . \qquad (6.29)$$

If $det_2(\Pi_1 T) \geq 0$, *or if* $M_1(T) \neq 0$, *then* S *is uniquely determined. If* $M_1(T) = 0$ *and if* $det_2(\Pi_1 T) < 0$, *then* $S \in SG_1 \cdot (\Pi_1 T)$ *iff* S *satisfies (6.29).*

b) (unoriented orbits) There exists a unique element $S \in V_1$ *such that*

$$\left.\begin{array}{ll} 1) & S \text{ is symmetric} \\ 2) & trS \geq 0 \\ 3) & det_2 S = |det_2(\Pi_1 T)| \\ 4) & G_1 \cdot S = G_1 \cdot (\Pi_1 T) \end{array}\right\} . \qquad (6.30)$$

Proof. First assume $M_1(T) \neq 0$. For $Q \in SO(2)$, (6.24) and (6.25) imply that

$$\left.\begin{array}{rl} QT:\hat{\underline{e}}_3 &= M_1(T)\sin(\phi - \theta_1) \\ tr_2(\Pi_1(QT)) &= M_1(T)\cos(\phi - \theta_1) \end{array}\right\} , \qquad (6.31)$$

where ϕ is the angle of rotation associated with Q, and

$$\cos\theta_1 = \frac{A + D}{M_1(T)} \;,\; \sin\theta_1 = \frac{(B - C)}{M_1(T)} .$$

Taking Q_1 to be specified by θ_1, taking $\sigma_1 = (Q_1, 1) \in G_1$ and $S = \Pi_1(\sigma_1 \cdot T)$ gives a representative satisfying (6.29). If $S_2 \in SG_1 \cdot \Pi_1 T$ satisfies (6.29), then property 3) implies $S_2 = Q(\Pi_1 T)$ for some $Q \in SO(2)$. Property 1) and (6.31) imply $S_2 = R_{m\pi} S$. So $tr(S_2) = (-1)^m trS$. Property 2) then implies

$S_2 = S$, so S is uniquely specified by (6.29) for this case. Now let $M_1(T) = 0$. Then $T_{11} = -T_{22} = A$, $T_{12} = T_{21} = B$. Hence, $\det_2(\Pi_1 T) \le 0$. If $\det_2(\Pi_1 T) = 0$, then $S = 0$ is the trivially unique element satisfying (6.29). If $\det_2(\Pi_1 T) < 0$, equations (6.24) through (6.26) imply that (6.29) hold for all $S = Q(\Pi_1 T)$, $Q \in SO(2)$.

b) If $\det_2(\Pi_1 T) \ge 0$, let S be the uniquely determined element in V_1 satisfying (6.29). If $\det_2(\Pi_1 T) \le 0$, $\det_2(\Pi_1(JT)) \ge 0$. So take S to be the uniquely defined element in V_1 which satisfies (6.29) for $\Pi_1(JT)$. In either case, S satisfies the first three properties of (6.30), with

$$\mathrm{tr} S = \left\{ \begin{array}{ll} M_1(T) & \text{if } \det_2(\Pi_1 T) \ge 0 \\ M_2(T) & \text{if } \det_2(\Pi_1 T) \le 0 \quad . \end{array} \right\} \tag{6.32}$$

Recognize that when $\det_2(\Pi_1 T) = 0$, $M_1(T) = M_2(T)$, and that the representative constructed by the two procedures coincide. Since

$$G_1 \cdot \Pi_1 T = SG_1 \cdot \Pi_1 T \cup SG_1(J,1) \cdot \Pi_1 T \;, \tag{6.33}$$

since $QJ = JQ^T$ for all $Q \in SO(2)$, and since $J(\Pi_1 T) = \Pi_1(JT)$, $SG_1(J,1) \cdot \Pi_1 T = SG_1(J,1) \cdot S$. Hence (6.33) and the fourth property of (6.29) imply the last property of (6.30). ∎

We obtain the following representations for the orbits of $\Pi_1 T$ under SG and G in terms of diagonal matrices. We call them oriented and unoriented orbits for $\Pi_1 T$. We justify the names below.

6.16 Proposition. *Let $T \in L(E^3)$. Let $M_1(T)$ and $M_2(T)$ be given by (6.28). Let SG be the component of G which contains the identity.*

a) (oriented orbits) There exists a unique element $N \in V_1$ which satisfies the following properties:

$$\left.\begin{array}{ll} 1) & N \text{ is diagonal} \\ 2) & \operatorname{tr} N \equiv 2a = M_1(T) \geq 0 \\ 3) & -\operatorname{tr}(JN) \equiv 2d \geq 0, \text{ where} \\ & \quad d^2 = a^2 - \det_2(\Pi_1 T) \\ 4) & SG\cdot N = SG\cdot\Pi_1 T\,. \end{array}\right\} \qquad (6.34)$$

b) (unoriented orbits) There is a unique element $N \in V_1$ *which satisfies the following properties:*

$$\left.\begin{array}{ll} 1) & N \text{ is diagonal} \\ 2) & \operatorname{tr} N \equiv 2a \geq 0, \text{ for} \\ & 2a = \begin{cases} M_1(T) & \text{if } \det_2(\Pi_1 T) \geq 0 \\ M_2(T) & \text{if } \det_2(\Pi_1 T) \geq 0 \end{cases} \\ 3) & -\operatorname{tr}(JN) \equiv 2d \geq 0, \text{ where} \\ & \quad d^2 = a^2 - |\det_2(\Pi_1 T)| \\ 4) & G\cdot N = G\cdot\Pi_1 T\,. \end{array}\right\} \qquad (6.35)$$

Proof.

a) By Lemma 6.15, there is an element $S \in V_1$ satisfying (6.29). Moreover, all such elements have the same trace and two-determinant. Lemma 6.14 then gives a uniquely determined element $N \in V_1$ which satisfies (6.23). Hence N satisfies the first three properties of (6.34), and

$$SG_1 S\Gamma\cdot N = SG_1 S\Gamma\cdot S. \qquad (6.36)$$

Since $QJ\Sigma_3 = J\Sigma_3 Q^T$ for all $Q \in SO(2)$,

$$SG_1 S\Gamma\cdot L = S\Gamma SG_1\cdot L \qquad (6.37)$$

for all $L \in L(3)$. Since $SG = SG_1 S\Gamma$, (6.36), (6.37), and property 4) of (6.29) give the last property of (6.34).

b) Let S be the unique element satisfying 96.30) by Lemma 6.15. By Lemma 6.14 and (6.32), we obtain a uniquely determined element $N \in V_1$ which satisfies (6.23). Hence, N satisfies the first three properties of (6.35), and

$$G_1\Gamma\cdot N = G_1\Gamma\cdot S. \tag{6.38}$$

The commutation relations among SO(2), J and Σ_3 imply

$$G_1\Gamma\cdot L = \Gamma G_1\cdot L \tag{6.39}$$

for all $L \in L(3)$. Since $G = G_1\Gamma$, (6.38), (6.39), and the fourth property of (6.30) give the last property of (6.35). ∎

Proposition 6.16 allows us to associate with a perturbing load a canonical forms which may preserve orientation or not, depending upon whether we use the orbits of SG or G to produce them.

6.17 Proposition. *Let $\bar{\xi} \in K$. Let $k(\bar{\xi},\gamma_0)$ be given by (3.32).*

a) Then there is at least one $g \in$ SG for which $N = \Pi_1(k(g^{-1}\bar{\xi},\gamma_0))$ satisfies (6.34). Moreover, all such elements in SG produce the same representative N in V_1.

b) There is at least one element $h \in$ G for which $N = \Pi_1(k(h^{-1}\bar{\xi},\gamma_0))$ satisfies (6.35). Moreover, all such elements in SG produce the same representative in V_1.

Proof. If $g \in G$, (3.32) implies

$$g^{-1}\cdot k(\bar{\xi},\gamma_0) = k(g^{-1}\bar{\xi},\gamma_0). \tag{6.40}$$

Taking $T = k(\bar{\xi},\gamma_0)$ and taking $g \in$ SG or G in turn, Proposition 6.16 and (6.40) imply the proposition. ∎

6.18 Definition. *Let $\bar{\xi} \in K$. If $g \in$ G (respectively SG), and if $\Pi_1(k(g^{-1}\bar{\xi},\gamma_0))$ satisfies (6.35) (respectively, (6.34)), call $g^{-1}\bar{\xi}$ a canonical form (respectively an oriented canonical form) for $\bar{\xi}$. If $\Pi_1(k(\bar{\xi},\gamma_0))$ satisfies (6.34) or (6.35), say $\bar{\xi}$ is in a canonical form.* ∎

Proposition 6.17 asserts that suitable rotations about the $\underline{e}_3$ axis and the centerline of the rod will transform a perturbing load into an oriented canonical form. Moreover, all such forms possess the same invariants given by (6.34). Possibly by reflecting the load along the $\underline{e}_1$ or $\underline{e}_3$ axis, we can transform the load into an unoriented canonical form. Hence the terminology involving orientation. Oriented canonical forms are convenient geometrically and physically. Unoriented forms will be more convenient mathematically.

The following lemma shows that to solve Problem 5.20, we may regard $\bar{\xi}$ as in a canonical form.

6.19 Lemma. *Let $\bar{\xi} \in \mathcal{K}$. Let $\xi = g_0^{-1}\bar{\xi}$ be a canonical form (oriented canonical form) for $\bar{\xi}$. The set of critical points solving Problem 5.20 for $\bar{\xi}$ are in bijective correspondence with the set solving Problem 5.20 for ξ.*

Proof. We prove the lemma for the case where ξ is not oriented. The proof for oriented canonical forms is a special case. Let $g_0 = \sigma_0\bar{g}_0 \in G \approx G_1\Gamma$. Then (5.31) implies

$$f(p,\bar{\xi},\sigma) = f(p,g_0\xi,\sigma) = f(p,\xi,\mathrm{Ad}_{\bar{g}_0^{-1}}(\sigma_0^{-1}\sigma)), \tag{6.41}$$

where $\mathrm{Ad}_{\bar{g}_0^{-1}}(\sigma_0^{-1}\sigma) \in G_1$ is given by (4.29). By Lemma 4.29, for $\sigma \in G_1$, $\mathrm{Ad}_{\bar{g}_0^{-1}}(\sigma_0^{-1}\sigma)$ defines a diffeomorphism of G_1. Hence, the chain rule implies that $\sigma \in \mathrm{critf}(p,\xi,\cdot)$ iff $\mathrm{Ad}_{\bar{g}_0^{-1}}(\sigma_0^{-1}\sigma) \in \mathrm{critf}(p,\bar{\xi},\cdot)$. Finally, Lemma 4.29 implies that if $\sigma_0 \in SG_1$, the diffeomorphism leaves the components of G_1 invariant. If $\sigma_0 \in SG_1(J,1)$, then the diffeomorphism maps one component of G_1 onto the other. ∎

In this development the use of Lemma 6.19 will parallel the use of daSilva's Theorem in Stoppelli's Problem (see [1], pp. 301).

We classify the perturbing loads by classifying the canonical forms in terms of the invariants (6.31) (See [31] §7). We present oriented and unoriented classifications separately.

6.20 Theorem. *Let $\bar{\xi} \in K$ be in canonical form. Let $k(\bar{\xi},\gamma_0) = T$, let $trT = 2a$ and $tr(JT) = -2d$.*

a) Suppose $\bar{\xi}$ is an oriented canonical form. Then depending upon a and d, T falls into one of three categories which determines $S_{\bar{\xi}}$.

category	*condition*	*prototypical T*
α	$a > 0,\ d > 0$	$\begin{pmatrix} a+d & 0 & r \\ 0 & a-d & s \\ g & h & 0 \end{pmatrix}$,
α_1	$a > d > 0$	
α_2	$a = d > 0$	
α_3	$d > a > 0$	
β		
β_1	$a > 0,\ d = 0$	$\begin{pmatrix} a & 0 & r \\ 0 & a & s \\ g & h & 0 \end{pmatrix}$,
β_2	$a = 0,\ d > 0$	$\begin{pmatrix} d & 0 & r \\ 0 & -d & s \\ g & h & 0 \end{pmatrix}$,
γ	$a = 0 = d$	$\begin{pmatrix} 0 & 0 & r \\ 0 & 0 & s \\ g & h & 0 \end{pmatrix}$.

category	$S_{\bar{\xi}}$
α	$\{1, (R_\pi,1), (J,1), (R_\pi J,1)\}$, four points in G_1.
β_1	$\{1, (R_\pi,1)\} \cup SG_1(J,1)$,
β_2	$SG_1 \cup \{(J,1), (R_\pi J,1)\}$, a circle and two points in G_1.
γ	G_1 two circles.

b) Suppose $\bar{\xi}$ is an unoriented canonical form. then depending upon a and d, T falls into one of three categories which specifies $S_{\bar{\xi}}$:

category	condition	prototypical T
α	$a \geq d > 0$	
α_1	$a > d > 0$	$\begin{pmatrix} a+d & 0 & r \\ 0 & a-d & s \\ g & h & 0 \end{pmatrix}$,
α_2	$a = d > 0$	$\begin{pmatrix} 2a & 0 & r \\ 0 & 0 & s \\ g & h & 0 \end{pmatrix}$,
β		
β_1	$a > 0,\ d = 0$	$\begin{pmatrix} a & 0 & r \\ 0 & a & s \\ g & h & 0 \end{pmatrix}$,
γ	$a = 0 = d$	$\begin{pmatrix} 0 & 0 & r \\ 0 & 0 & s \\ g & h & 0 \end{pmatrix}$.

The specification of $S_{\bar{\xi}}$ is as given for the corresponding subclassification for the oriented case.

c) All canonical forms for a given load in $\mathcal{K}$ fall into a single category, and their critical manifolds for the forms coincide.

Proof.

a) and b). We establish the proposition for the oriented case. The unoriented case follows by the same argument upon using (6.35) and restricting d by $a \geq d \geq 0$. By the hypothesis and (6.34) $\Pi_1 T$ is diagonal, and a and d are not negative. By (6.19) and (6.40),

$$S_{\bar{\xi}} = \{ \sigma \in G_1 \mid \sigma^{-1} \cdot T : \hat{\underline{e}}_3 = 0 \}. \tag{6.42}$$

If $\sigma = Q,1) \in G_1$, the hypothesis, (6.24), (6.26), and the commutation relations between SO(2) and J imply

$$(\sigma^{-1}\cdot T):\hat{\underline{e}}_3 = \begin{cases} -2ay & \text{if } Q = R_\phi \in SO(2) \\ 2dy & \text{if } Q = R_\phi J \quad , \end{cases} \tag{6.43}$$

where $y = \sin\phi$. The three categories exhaust the possibilities. Since $G_1 = SG_1 \cup SG_1(J,1)$, we may use (6.43) to compute (6.42) for each category in turn to produce the conclusion for $S_{\bar{\xi}}$.

c) Let ξ and $\bar{\xi}$ be two (oriented or unoriented) canonical forms for $\xi \in \mathcal{K}$. By (6.40), Definition 6.18, and Proposition 6.17, $k(\xi,\gamma_0)$ and $k(\bar{\xi},\gamma_0)$ determine the same values for a and d. Hence, the conclusion follows from parts a) and b). ∎

Remarks.

1. The three categories provide a primary classification for the perturbing loads. The subcategories herald the need for a finer classification which we encounter when we study the full problem at $p = p_0$, or degenerate cases in the orbit breaking problem for $p < p_0$.

2. It is convenient to view the oriented and unoriented classifications schematically in terms of the parameters a and d. We present these diagrams in Figure 6.3.1. As the figure suggests, when we relax the condition of orientation in the criterion for equivalence, we identify the points (a,d) and (d,a) in the oriented diagram when $d \geq a$. Hence, we achieve the unoriented diagram by folding the oriented diagram along the $a = d$ line, identifying the classes α_1 with α_3, and β_2 with β_1.

3. As indicated previously, the oriented classification is convenient from a geometric or physical perspective. The unoriented classification is more convenient for the mathematical analysis.

The strategy for solving Problem 5.20 will be to solve the problem for each class of canonical forms in Theorem 6.20. We will then invoke Lemma 6.19 to solve the problem of interest.

We close this subsection by illustrating loads of each category arising in Theorem 6.20. We classify the loads presented as illustrations at the end of Section III-2.

6.21 Example. *A Parallel Force Distribution.* Consider the load of Example 3.19:

$$\bar{\xi} = (\mu,0), \quad \mu(S) \equiv \pi R^3(\underline{e}_1 \otimes \underline{e}_1),$$

which is illustrated in Figure 3.1.1. By (3.32), $T_1 = k(\bar{\xi},\gamma_0) = \pi R^3(\underline{e}_2 \otimes \underline{e}_2)$; hence, $\bar{\xi}$ is in canonical form. Since $2a = 2d = \pi R^3$, $\bar{\xi}$ is of type α_2, and $S_{\bar{\xi}}$ consists of four distinct points in G_1. ∎

6.22 Example. *A Homogeneous Non-Parallel Force Distribution.* Consider the load of Example 3.20 a):

$$\bar{\xi} = (\mu,0), \quad \mu(S) \equiv \pi R^3(\underline{e}_1 \otimes \underline{e}_1 - \underline{e}_2 \otimes \underline{e}_2),$$

which is illustrated in Figure 3.1.2. By (3.32) $T_2 = k(\bar{\xi},\gamma_0) = \pi R^3(\underline{e}_1 \otimes \underline{e}_1 - \underline{e}_2 \otimes \underline{e}_2)$, so $\bar{\xi}$ is an oriented canonical form. Since $2a = 0$, and $2b = \pi R^3$, $\bar{\xi}$ is of type β_2, and $S_{\bar{\xi}}$ consists of two points and one circle in G_1. In contrast, an unoriented canonical form which represents $\bar{\xi}$ is the load presented in Example 3.21 b):

$$\tilde{\xi} = (\mu,0), \quad \mu(S) \equiv \pi R^3(\underline{e}_1 \otimes \underline{e}_1 + \underline{e}_2 \otimes \underline{e}_2),$$

which is illustrated in Figure 3.1.3. $S_{\tilde{\xi}}$ is of class β_1. By examining Examples 3.12 and 3.13 and the figures, we recognize how we may erect $\bar{\xi}$ from $\tilde{\xi}$ by reflecting the three-dimensional force distribution producing $\tilde{\xi}$ along the $\underline{e}_1$ axis. ∎

6.23 Example. *An Inhomogeneous, Non-Parallel Force Distribution.* Modify Examples 3.12 and 3.20 in the following way. Let $\underline{X} = (X^1,X^2,S)$. As a three-dimensional force distribution, take $l = (0,\tau)$,

$$\tau(\underline{X}) = \begin{cases} X^1 \underline{e}_1 - \dfrac{\pi X^2}{2} \sin(\pi S)\underline{e}_2, & \underline{X} \in \partial B(S),\ 0 < S < 1 \\ 0 & S = 0,1. \end{cases}$$

Averaging over the cross sections of the rod in the configuration γ_0 gives

$$\bar{\xi} = (\mu, 0), \quad \mu(S) = \pi R^3 \left(\underline{e}_1 \otimes \underline{e}_1 - \frac{\pi \sin(\pi S)}{2} \underline{e}_2 \otimes \underline{e}_2 \right).$$

By (3.32), $T_3 = k(\bar{\xi}, \gamma_0) = \pi R^3(\underline{e}_1 \otimes \underline{e}_1 + \underline{e}_2 \otimes \underline{e}_2)$, so $\bar{\xi}$ is in canonical form, $\bar{\xi}$ is of type β_1, and $S_{\bar{\xi}}$ consists of two distinct points and one circle in G_1. ■

6.24 Example. *A Constant Gravitational Force Distribution.* Consider the load of Example 3.21:

$$\bar{\xi} = (\mu, 0), \quad \mu(S) = -\pi R^2 (1 - S)(\underline{e}_2 \otimes \underline{e}_3),$$

which is illustrated in Figure 3.1.4. By (3.32), $T_4 = k(\bar{\xi}, \gamma_0) = -(1/2)\pi R^2(\underline{e}_2 \otimes \underline{e}_3)$, so $\bar{\xi}$ is in canonical form. Since $a = d = 0$, $\bar{\xi}$ is of type γ, and $S_{\bar{\xi}} = G_1$. ■

DIAGRAMS FOR THE CLASSIFICATION THEOREM 6.20

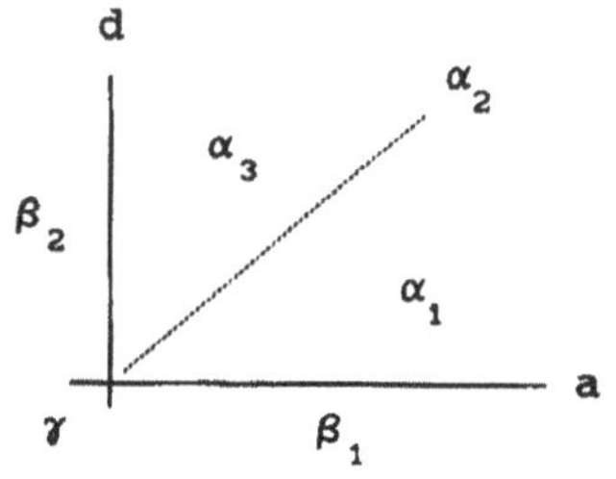

Oriented Diagram

$a \geq 0,\ d \geq 0$

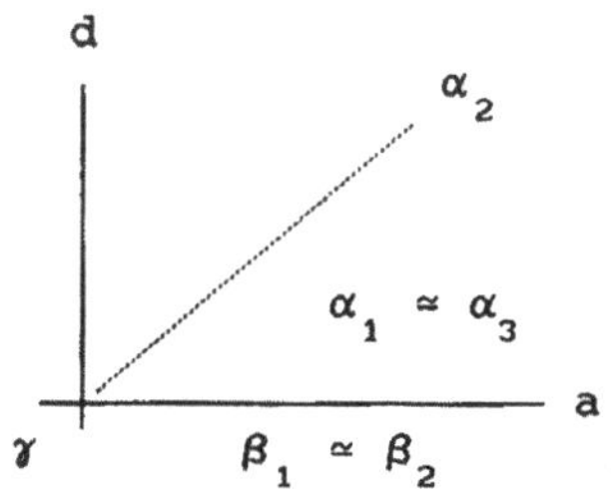

Unoriented Diagram

$a \geq d \geq 0$

Figure 6.3.1

VII. THE RESULTS FOR THE BIFURCATION PROBLEM

In this section we determine how the trivially equilibrating orbit $G\gamma_0$ for the unperturbed Problem 5.20 alters when we perturb ξ_0 by each type of load arising from the classification of Theorem 6.20. The analysis parallels that used on Stoppelli's Problem for type 0 and type 1 loads in [1], modified in the manner suggested in [3] to include the presence of the symmetry group Γ.

VII.1. *The Reduction of ℓ and its Analysis*

We will use the Morse theory to examine when the zeroth and first order terms in (6.7) determine the critical points for $\ell(p,\tau,\cdot;\bar{\xi})$ on G_1. We begin with the following result from the Morse theory.

7.1 Definition.

a) *If V is a vector space, if D is a quadratic form on V, call D* nondegenerate *on V if for $v \in V$, $D(v) = 0$ implies $v = 0$.*

b) *Let f be a C^2 function on a smooth manifold function M.*

(1) *Call $x \in M$* a nondegenerate critical point *for f if $Tf(x) = 0$, and if $T^2f(x)$ determines a quadratic form on T_xM which is nondegenerate.*

(2) *Call f* a Morse function *if every critical point of f is nondegenerate.*

c) *If x is a nondegenerate critical point for f, call the* index for f at x *the dimension of the largest subspace of T_xM on which $T^2f(x)$ is negative definite.* ∎

7.2 Lemma. *Let M be a smooth, compact manifold, and let f be a smooth (C^∞) Morse function on M. Then for $g \in C^\infty(M,\mathbb{R})$ sufficiently close to f in the C^2 sense,*

a) *g is a Morse function,*

b) *the critical points for g are near and in bijective correspondence with those for f, and*

c) *corresponding critical points have the same index.*

Proof. By theorem ([16] p. 79), the set of Morse functions is open in $C^\infty(M,\mathbb{R})$. Hence, there is a neighborhood $\mathcal{N}(f)$ of f in $C^\infty(M,\mathbb{R})$, such that $g \in \mathcal{N}(f)$ implies that g is a Morse function. An examination of the proof of the theorem indicates that the neighborhood may be taken from the Whitney C^2 topology on $C^\infty(M,\mathbb{R})$. Parts b) and c) follow from the fact that a Morse function is locally stable about a critical point ([25], pp. 15, 22).

Lemma 7.2 allows us to determine when the zeroth and first order terms in (6.7) determine nondegenerate critical points for $\ell(p,\tau,\cdot;\bar{\xi})$. In the light of Theorem 6.20 we formulate the proposition this way.

7.3 Proposition. *Let* $\bar{\xi} \in \mathcal{K}$, *let* $\bar{p} < p_0$, *and let* p *and* τ *satisfy the conditions of Problem 5.20.*

a) If $A_{\bar{\xi}}$ *is a Morse function on a component of* G_1, *then for* $\tau > 0$ *sufficiently small,*

(1) $\ell(p,\tau,\cdot;\bar{\xi})$ *is a Morse function on that component,*

(2) the critical points for $A_{\bar{\xi}}$ *on the component are near and in bijective correspondence with those for* $\ell(p,\tau,\cdot;\bar{\xi})$,

(3) the indices for the corresponding critical points for the two functions are the negatives of each other.

b) Let $u_1(p,\sigma;\bar{\xi})$ *be given by (6.2). If* $A_{\bar{\xi}}$ *is constant on a component of* G_1, *and if* $K(L_p u_1(p,\cdot;\bar{\xi}),u_1(p,\cdot;\bar{\xi}))$ *is a Morse function on that component of* G_1, *then for* $\tau > 0$ *sufficiently small,*

(1) $\ell(p,\tau,\cdot;\bar{\xi})$ *is a Morse function on that component of* G_1,

(2) the critical points for $K(L_p u_1(p,\cdot;\bar{\xi}),u_1(p,\cdot;\bar{\xi}))$ *are near and in bijective correspondence with those for* $\ell(p,\tau,\cdot;\bar{\xi})$ *on the component,*

(3) the indices of the corresponding critical points for the two functions are the negatives of each other.

Proof.

a) Let $h(p,\tau,\sigma;\bar{\xi})$ be given by (6.5). By the Mean Value Theorem, it is a smooth function of its arguments. By (6.7),

$$h(p,\tau,\sigma;\bar{\xi}) = - A_{\bar{\xi}}(\sigma) + O(\tau). \tag{7.1}$$

Since $A_{\bar{\xi}}(\cdot)$ is a Morse function by hypothesis, Lemma 7.2 implies that for τ suitably small, $h(p,\tau,\cdot;\bar{\xi})$ is also a Morse function, and that corresponding critical points for $A_{\bar{\xi}}(\cdot)$ and $h(p,\tau,\cdot;\bar{\xi})$ have indices which are negatives of one another. Proposition 6.5 and Lemma 7.2 then imply part a).

b) Let $A_{\bar{\xi}}(\sigma) \equiv c_1$ on the component in question. Define $h_2(p,\tau,\sigma)$ by

$$h(p,\tau,\sigma;\bar{\xi}) = - c_1 + (\tau/2)h_2(p,\tau,\sigma;\bar{\xi}) . \tag{7.2}$$

By (6.7) and the Taylor Theorem h_2 is a smooth function of its arguments, and

$$h_2(p,\tau,\sigma;\bar{\xi}) = - K(L_p u_1(p,\bar{\sigma};\xi),u_1(p,\bar{\sigma};\xi)) + O(\tau). \tag{7.3}$$

By hypothesis, the first term on the right hand side of (7.3) is a Morse function on G_1, depending parametrically on p. So the argument of part a) applied to (7.2) and (7.3), implies part b) ∎

We can now use Theorem 6.20 and Proposition 7.3 to resolve Problem 5.20.

Remark. The conclusions of Lemma 7.2 and Proposition 7.3 may be expressed in terms of the gradient of $\ell(p,\tau,\cdot;\bar{\xi})$ using the theory of singularities. Since the factors involved are assumed to be Morse functions, nothing is gained by the reformulation. However, when we pursue Problem 5.25, it will be important to so reformulate these propositions. We will reserve this reformulation for that time.

Remark. To anticipate the resolution of Problem 5.25 it is instructive to examine the previous development in terms of the theory of singularities and (6.9). We make the following observations.

1. In Lemma 7.2 let $H(x) = Tf(x)$. Then the lemma says about each singular point x_0 for f, H is contact equivalent ([23], p.166) to $n_1(x) = c_1 x$, $c_1 \neq 0$, and that the codimension of n_1 is zero.

2. If we let

$$g(\sigma) = -\frac{\partial}{\partial\sigma} A_{\bar{\xi}}(\sigma)$$

then (6.9) implies

$$h(p,\tau,\sigma) = g(\sigma) + \tau p(p,\tau,\sigma) .$$

Proposition 7.3a: may then be interpreted as recognizing

$$g(\sigma) \simeq n_1(\sigma) = -\delta\sigma ,$$

that n_1 has (contact) codimension zero, and that $h(p,\tau,\sigma)$ is an unfolding of a codimension zero germ. Consequently, there is a change of variables $\Sigma(p,\tau,\sigma)$ of G_1 which takes $n_1(\sigma)$ into $h(p\ \tau,\sigma)$, modulo a sign-preserving scale factor.

3. The hypothesis of Proposition 7.3b) and (6.9) imply that

$$h(p,\tau,\sigma) = \tau g_2(p,\sigma;) + \tau^2 q(p,\tau,\sigma) ,$$

where

$$g_2(p,\sigma;\xi) = -\frac{\partial}{\partial\sigma} \langle K(L_{p+\lambda}u_1(p,\sigma;\xi),u_1(p,\sigma;\xi))\rangle .$$

Hence, for $\tau > 0$, the zeros of $h(p,\tau,\cdot)$ are in bijective correspondence with those of $g_2(p,\cdot) + \tau q(p,\tau,\cdot)$. Assuming that $K(L_{p+\lambda}u_1(p,\sigma;\xi),u_1(p,\sigma;\xi))$ is equivalent to assuming $g_2(p,\sigma;\xi)$ is
contact equivalent to $n_1(\sigma)$. Hence, the conclusions of the previous remark follow for this case.

Proof. Parts a) and b) follow from the compactness of M and the fact that Morse functions are locally stable ([25], p. 22). Part c) follows in the manner of [26], p. 119. ■

VII.2. *Perturbations of Class α*

In this subsection we resolve Problem 5.20 for the perturbations of class α of Theorem 6.20. We begin by determining when the first order term $A_{\bar{\xi}}$ of (6.6) can have nondegenerate critical points on a component of G_1 and what the index of the critical point will be.

7.4 Lemma. *Let $\bar{\xi} \in \mathcal{K}$ and $\sigma \in S_{\bar{\xi}}$. Then*

$$\left(A_{\bar{\xi}}(\sigma)\right)_{\sigma\sigma} = - tr(k(\sigma^{-1}\bar{\xi},\gamma_0)). \tag{7.4}$$

Proof. Let $\sigma = (Q,1) \in G_1$, $Q \in O(2)$. By (6.6), (6.15), and (6.40),

$$\left(A_{\bar{\xi}}(\sigma)\right)_{\sigma} = \hat{\underline{e}}_3 : k(\sigma^{-1}\bar{\xi},\gamma_0) \ .$$

If $\sigma \in S_{\bar{\xi}}$, we may evaluate the second derivative of $A_{\bar{\xi}}$ at σ without formally invoking the connection on G_1. The chain rule gives

$$\left(A_{\xi}(\sigma)\right)_{\sigma\sigma} = (\hat{\underline{e}}_3\hat{\underline{e}}_3) : k(\sigma^{-1}\bar{\xi},\gamma_0) \ ,$$

where the juxtaposition of the $\hat{\underline{e}}_3$ denotes a composition.

Since
$$(\hat{\underline{e}}_3\hat{\underline{e}}_3)\underline{e}_j = \begin{cases} -\underline{e}_j & j = 1,\ 2 \\ 0 & j = 3 \end{cases} \ ,$$

equation (7.4) follows. ∎

7.5 Proposition. *Let $\bar{\xi} \in \mathcal{K}$ be a canonical form. Let σ be a nondegenerate critical point for $A_{\bar{\xi}}$ on G_1. Then the index of σ for $A_{\bar{\xi}}$ is either 0 or 1.*

Proof. Evaluate (7.4) for each of the cases in Theorem 6.20 which admit nondegenerate critical points. Setting $T = k(\bar{\xi},\gamma_0)$ we obtain the following results.

class α: $\sigma \in \{1, (R_\pi,1), (J,1), (R_\pi J,1)\} \subseteq G_1$.

$$\operatorname{trk}(\sigma^{-1}\bar{\xi},\gamma_0) = \left\{\begin{array}{cc} & \underline{\sigma} \\ \operatorname{tr}(T) & 1 \\ -\operatorname{tr}(T) & (R_\pi,1) \\ \operatorname{tr}(JT) & (J,1) \\ -\operatorname{tr}(JT) & (R_\pi J,1) \end{array}\right\}. \tag{7.5}$$

class β_1: $\sigma \in \{1, (R_\pi,1)\} \subseteq G_1$.

$$\operatorname{trk}(\sigma^{-1}\bar{\xi},\gamma_0) = \left\{\begin{array}{cc} & \underline{\sigma} \\ \operatorname{tr}(T) & 1 \\ -\operatorname{tr}(T) & (R_\pi,1) \end{array}\right\}. \tag{7.6}$$

class β_2: $\sigma \in \{(J,1), (R_\pi J,1)\} \subseteq G_1$.

$$\operatorname{trk}(\sigma^{-1}\bar{\xi},\gamma_0) = \left\{\begin{array}{cc} & \underline{\sigma} \\ \operatorname{tr}(JT) & (J,1) \\ -\operatorname{tr}(JT) & (JR_\pi,1) \end{array}\right\}. \tag{7.7}$$

Equation (7.4) and the above results imply that the index for $A_{\bar{\xi}}$ at σ is either 0 or 1, depending upon σ and the signs of tr(T) and tr(JT). ∎

Remarks.

1. Since G_1 is a one-dimensional manifold, a nondegenerate critical point for $A_{\bar{\xi}}$ on G_1 will be either a minimum point or a maximum point.

2. Proposition 7.5 provides the condition ($\mathfrak{B}$) of [3], p. 221, for Problem 5.20.

3. Equation (6.40) and the invariance of the trace under a similarity transformation imply that the conclusion of Proposition 7.5 is independent of the choice of canonical form for $\bar{\xi}$.

Proposition 7.5 resolves Problem 5.20 for perturbations of the class α of Theorem 6.20.

7.6 Theorem. *Let $\bar{p} < p_0$ and let p, τ satisfy the conditions of Problem 5.20. Let $\bar{\xi}$ be a perturbation of class α. For $\tau > 0$ sufficiently small, in the neighborhood of $G\gamma_0$ in M_1 there are four configurations $\{\gamma_i(p,\tau) \mid i = 1, 2, 3, 4\}$ which equilibrate the load $\xi = p\xi_3+\tau\bar{\xi}$. Two of them are stable and two unstable with respect to spatial perturbation. One configuration of each stability can be obtained by a continuous deformation from γ_0. As τ approaches zero, $\gamma_i(p,\tau)$ approaches $\sigma_i\gamma_0$, where the $\sigma_i \in S_{\bar{\xi}} \subseteq G_1$.*

Proof. By (7.4), (7.5), Theorem 6.20, and Proposition 7.5, $A_{\bar{\xi}}$ is a Morse function with four critical points, two of which have index equal to one, two have index equal to zero. Since G_1 is one-dimensional, the indices correspond to maxima and minima for $A_{\bar{\xi}}$, respectively. By (7.5) a critical point of each stability lies on each component of G_1. Theorem 5.18, (5.32), and Proposition 7.3 then give the first three conclusions. The last conclusion follows from the characterization of Morse functions by a condition of transversality ([16], p. 63), and the isotopy property of a transversal intersection ([32], p. 51). ∎

Thus, we may resolve Problem 4.25 in the following way for the case of a one-parameter perturbation of class α. For $p < p_0$, for $\tau > 0$, there will be four equilibrating configurations for the perturbed problem nearby the orbit $G\gamma_0$ of trivially equilibrating configurations for the unperturbed problem. These equilibria may be viewed as four families "branching" from four distinct configurations on the orbit $G\gamma_0$ for the unperturbed problem.

Using (6.5) and (7.1), we may visualize the branching as the graphs of four paths $\sigma_i(p,\tau)$, $1 \le i \le 4$, in $G_1 \times \mathbb{R}$ which depend parametrically on p. The curves emerge from four distinct elements $\sigma_i(p,0) = \sigma_i \in S_{\bar{\xi}}$ at $\tau = 0$, elements which are independent of p. Those $\sigma_i(p,\tau)$ lying on SG_1 represent configurations which may be gained from γ_0 by a continuous deformation. Those lying on $SG_1(J,1)$ require an inversion of the cross sections of the rod.

Finally, since $A_{\bar{\xi}}$ depends only on $\bar{\xi}$, σ, and γ_0, these conclusions are *independent* of the particular material comprising the body, so long as it satisfies the hypotheses of the problem. A perturbation of class α so breaks the symmetry of ξ_0 that changing the material comprising the body will not affect the breaking of the orbit $G\gamma_0$ for the unperturbed problem.

We close the subsection by illustrating the branching conclusions of Theorem 7.6 and the subsequent remarks for the class α perturbing load of Example 6.21. As in Figure 3.1.1 let R_ϕ be a rotation about $\underline{e}_3$ through an angle ϕ, let $\sigma_\phi = (R_\phi,1) \in G_1$, and let $\sigma_\phi\gamma_0$ be the rotated trivial configuration. By Lemma 4.23 and (4.14), $\sigma_\phi\gamma_0$ is a stress-free configuration for the material comprising the rod. So it is an equilibrating configuration for the load $\xi_0 + \tau\bar{\xi}$ if and only if the torque produced by $\bar{\xi}$ in the configuration $\sigma_\phi\gamma_0$ can be balanced by the constraint forces. The figure suggests that such a balance will occur when $\phi = 0$ and $\phi = \pi$, and that $\sigma_0 = 1$ will produce a stable equilibrium for the perturbed load, while σ_π will produce an unstable one.

Theorem 7.6 allows us to complete the description for $\bar{\xi}$ suggested by the figure. By (6.40) and the remark following Lemma 6.10, the unbalanced torque produced by $\bar{\xi}$ in the configuration $\sigma_\phi\gamma_0$ is

$$k(\sigma_\phi^{-1}\bar{\xi},\gamma_0):\hat{\underline{e}}_3 = \pi R^3 \sin\phi.$$

So balance occurs when $\phi = 0$ and $\phi = \pi$. But Theorem 6.20 also indicates that two other equilibria occur : at $\sigma = (J,1)$ and $\sigma = \sigma_\pi(J,1)$. Since $\operatorname{tr}k(\bar{\xi},\gamma_0) = \operatorname{tr}(Jk(\bar{\xi},\gamma_0)) = \pi R^3 > 0$, (7.1) and

(7.4) indicate that $\sigma_0 = 1$ and (J,1) are the stable equilibria for $h(p,0,\sigma)$ and σ_π and $(J,1)\sigma_\pi$ are the unstable ones.

Theorem 7.6 now asserts that the equilibria for $p\xi_3 + \tau\bar{\xi}$ in the vicinity of $G\gamma_0$ may be represented as distinct paths branching from these four elements of G_1. Figure 7.2.1 illustrates only those configurations which can be obtained from γ_0 by a continuous deformation of the rod. The two associated with (J,1) and σ_π(J,1) require an inversion of the cross sections of the rod.

Illustration of the Branching Theorem

Case α

Example 6.21 $\xi_1 = (\pi R^3(\underline{e}_1 \otimes \underline{e}_1),0)$

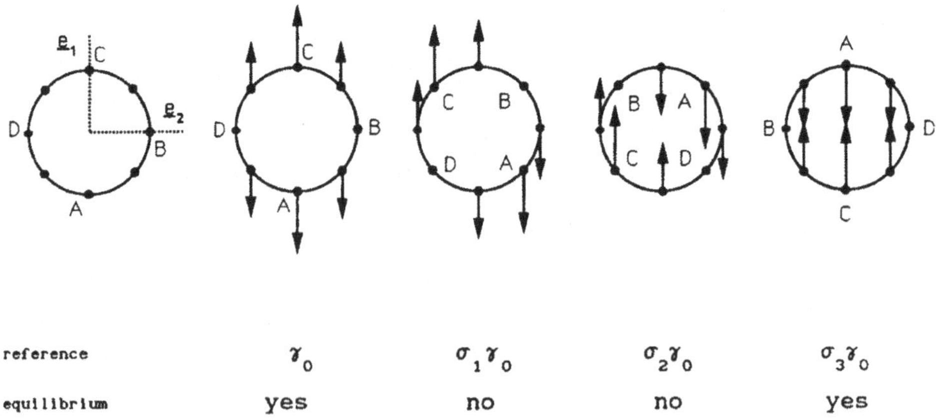

Figure 7.2.1

VII.3. *Nondegenerate Perturbations of Classes β and γ*

We now examine how the orbit of trivially equilibrating configurations alters when ξ_0 is perturbed by a load of class β or γ. As Theorem 6.20 indicates, in such cases $S_{\bar{\xi}}$ contains an entire component of G_1. In this section we examine situations where the terms in $\ell(p,\tau,\cdot;\bar{\xi})$ of order two in τ determine its critical points on such components when $\tau > 0$. We begin by examining the behavior of the second-order term of (6.5).

7.7 Notation. *If $\sigma \in G_1$, depending upon the component of G_1 in which σ resides (see Lemma 4.28), represent σ as $\sigma = (Q,1)$ or $\sigma = (QJ,1)$, for $Q \in SO(2)$. Let (x,y) specify Q as in Notation 6.11.* ∎

7.8 Lemma. *For $\sigma \in G_1$, $K(L_p u_1(p,\sigma;\bar{\xi}),u_1(p,\sigma;\bar{\xi}))$ is a polynomial of degree ≤ 2 in x and y.*

Proof. By (6.4),

$$L_p u_1(p,\sigma;\bar{\xi}) = \Pi_{2e}\beta(\sigma^{-1}\bar{\xi},\gamma_0). \tag{7.8}$$

Since $\sigma \in G_1$, (3.39) implies that the components of $\sigma^{-1}\bar{\xi} \in \mathcal{K}$ are polynomials of degree ≤ 1 in x and y with coefficients which may vary with $S \in I$. Since β is linear in its first variable, (7.8) implies that the components of $L_p u_1(p,\sigma;\bar{\xi})$ are polynomials of degree ≤ 1 in x and y whose coefficients are elements of $\mathcal{F}_{2e}(\gamma_0)$. Since L_p is a linear mapping which takes $U_2(\gamma_0)$ into $\mathcal{F}_{2e}(\gamma_0)$, it follows that the components of $u_1(p,\sigma;\bar{\xi})$ are polynomials of degree ≤ 1 in x and y whose coefficients are elements of $U_2(\gamma_0)$. As K is a bilinear form, the proposition follows. ∎

Remark. Lemma 5.5 and (7.8) imply that the coefficients of the polynomial in Lemma 7.8 are determined by $\bar{\xi}$ and the linearly elastic behavior about γ_0 of the material comprising the rod.

Lemma 7.8 allows us to apply the technique used in [1] to analyze when the second order term in (6.7) has nondegenerate critical points in G_1.

7.9 Proposition. *On a connected component of* G_1 *write*

$$\left.\begin{aligned} K(L_p u_1(p,\sigma;\bar{\xi}),u_1(p,\sigma;\bar{\xi})) &= \alpha_1 x^2 + 2\alpha_2 xy + \alpha_3 y^2 \\ &\quad + \alpha_4 x + \alpha_5 y + \alpha_6, \end{aligned}\right\} \tag{7.9}$$

where (x,y) *represents* σ *as in Notation 7.7. Let* $\sigma_0 = (Q_0,1) \in SG_1$ *be the left translation in* G_1 *which diagonalizes the principal part of (7.9):*

$$\left.\begin{aligned} Q(p,\sigma) &\equiv K(L_p u_1(p,\sigma_0\sigma;\bar{\xi}),u_1(p,\sigma_0\sigma;\bar{\xi})) \\ &= a_1 X^2 + a_3 Y^2 + a_4 X + a_5 Y + a_6 \end{aligned}\right\}, \tag{7.10}$$

where $(X,Y) = (x,y)Q_0^T$.

a) *Suppose that* $a_1 \neq a_3$. *Set (see [1], p. 323)*

$$\Delta = [(2(a_1 - a_3))^2 - a_4^2 - a_5^2]^3 - 108a_4^2 a_5^2 (a_1 - a_3)^2. \tag{7.11}$$

Then (7.9) is a Morse function on the component of G_1 *if and only if* $\Delta \neq 0$.

1) *If* $\Delta > 0$ *(7.9) has 4 critical points on the component, two of index 0, two of index 1.*

2) *If* $\Delta < 0$ *(7.9) has 2 critical points on the component, one of index 0, one of index 1.*

b) *If* $(a_1 - a_3) = 0$, *(7.9) is a Morse function if and only if* $\Delta \neq 0$. *In such a case (7.9) has two critical points on the component, one of index 0, one of index 1.*

Proof. Let $(a_3 - a_1) \neq 0$. Equating $Q_\sigma(p,\sigma)$ and $Q_{\sigma\sigma}(p,\sigma)$ to zero shows that $Q(p,\cdot)$ is a Morse function on G_1 if and only if $\Delta \neq 0$. (See [1], p. 323ff for a more comprehensive proof.) As in [1], p. 323, show that the projection Π of $S^1 \times \mathbb{R}^2$ onto $\mathbb{R}^2$ given by $\Pi(Q,a_4,a_5) = (a_4,a_5)$, restricted to

$$M = \{(Q,a_4,a_5) \mid Q_\sigma(p,\sigma) = 0,\ \sigma = (Q,1)\}$$

is a regular map at those points of M for which $\Delta \neq 0$. Consequently, the cardinality of $\Pi^{-1}(a_4,a_5) \cap M$ is constant on the components of (a_4,a_5)-space given by $\Delta < 0$ and $\Delta > 0$. Evaluation at specific choices of (a_4,a_5) gives parts 1) and 2) of a). When $a_1 = a_3$ the expressions for Q_σ and $Q_{\sigma\sigma}$ show $Q(p,\cdot)$ is a Morse function if and only if $(a_4,a_5) \neq (0,0)$, or that $\Delta \neq 0$, and that the number of critical points and their indices are as asserted. ∎

We now resolve Problem 5.20 for perturbations of class β and γ when the second-order term in (6.7) satisfies the condition $\Delta \neq 0$ on those components of G_1 on which $A_{\bar\xi}$ is constant.

7.10 Theorem. *Let $\bar\xi \in \mathcal{K}$ be a load of class β, let $\bar p < p_0$, and let p, τ satisfy the conditions of Problem 5.20. Let $Q(p,\sigma)$ and Δ be specified by (7.10) and (7.11), respectively.*

- *a) Let $(a_3 - a_1) \neq 0$. If $\Delta \neq 0$ on the component of G_1 on which $A_{\bar\xi}$ is constant, then for $\tau > 0$ sufficiently small, in a neighborhood of $G\gamma_0$ in $\mathcal{M}_1$ there are either four or six configurations which equilibrate the load $\xi = p\xi_3 + \tau\bar\xi$. The number depends upon the sign on Δ.*
- *b) Let $(a_3 - a_1) = 0$. If $\Delta \neq 0$ on the component of G_1 on which $A_{\bar\xi}$ is constant, then for $\tau > 0$ sufficiently small, in a neighborhood of $G\gamma_0$ in $\mathcal{M}_1$ there are four configurations which equilibrate the load $\xi = p\xi_3 + \tau\bar\xi$.*

Half of the configurations are stable with respect to spatial perturbations, half are unstable. Moreover, as τ approaches 0, each of the equilibrating configurations $\gamma_1(p,\tau)$ approaches $\sigma_1(p)\gamma_0$, where $\sigma_1(p) \in G_1$ is one of the critical points for $A_{\bar\xi}$ or $Q(p,\cdot)$, depending upon the component of G_1 on which it lies.

Proof. By Lemma 6.19 we may take $\bar\xi$ to be in canonical form. By Theorem 6.20 and 7.5, $A_{\bar\xi}$ is a Morse function on one component of G_1 with two critical points. One point is of index 1, the other is of index 0. By Theorem 7.6, the same conclusions hold for $\ell(p,\tau,\cdot;\bar\xi)$ on that component, provided $\tau > 0$ is sufficiently small. On the component on which $A_{\bar\xi}$ is constant, if $a_3 \neq a_1$, the hypothesis and Proposition 7.9 imply that $Q(p,\cdot)$. Hence, $K(L_p u_1(p,\cdot;\bar{\xi}),u_1(p,\cdot;\bar{\xi}))$ is a Morse function having either two or four critical points, depending upon the sign of Δ. Likewise, if $a_3 = a_1$, $Q(p,\cdot)$ is a Morse function with two critical points. Half of the critical points have index 1, half have index 0. Proposition 7.3 implies that the same conclusions hold for $\ell(p,\tau,\cdot;\bar{\xi})$ on that component, provided $\tau > 0$ is sufficiently small. Theorem 5.18 and (5.32) then give the first three conclusions. The last conclusion follows in the same manner as in Theorem 7.6 ■.

7.11 Theorem. *Let $\bar{\xi} \in \mathcal{K}$ be a load of class γ, let $\bar{p} < p_0$, and let p, τ satisfy the conditions of Problem 5.20. Let $Q(p,\sigma)$ and Δ be given by (7.10) and (7.11), respectively.*

a) Let $(a_1 - a_3) \neq 0$ If $\Delta \neq 0$ on each component of G_1, then for $\tau > 0$ sufficiently small, in a neighborhood of $G\gamma_0$ in $\mathcal{M}_1$ there are either four, six, or eight configurations equilibrating the load $\xi = p\xi_3 + \tau\bar{\xi}$. Half of the configurations are stable, half are unstable. The number depends upon the signs of Δ on the components of G_1.

b) Let $a_1 = a_3$. If $\Delta \neq 0$ on each component of G_1, then for $\tau > 0$ sufficiently small, in a neighborhood of $G\gamma_0$ in $\mathcal{M}_1$ there are four configurations equilibrating the load $\xi = p\xi_3 + \tau\bar{\xi}$. Half of the configurations are stable, half are unstable.

Moreover, as τ approaches 0, each of the equilibrating configurations $\gamma_1(p,\tau)$ approaches $\sigma_1(p)\gamma_0$, where $\sigma_1(p) \in G_1$ is one of the nondegenerate critical points for $Q(p,\cdot)$.

Proof. By Lemma 6.19 we may take $\bar{\xi}$ to be in canonical form. By Theorem 6.20, $A_{\bar{\xi}}$ is constant on each component of G_1. If $a_1 - a_3 \neq 0$, Proposition 7.9 and the hypotheses, imply that $K(L_p u_1(p,\sigma;\bar{\xi}),u_1(p,\sigma;\bar{\xi}))$ is a Morse function with either two or four critical points on each component, depending upon the sign of Δ. If $a_1 - a_3 = 0$, the function is Morse with two critical points on each component of G_1. Half of the critical points have index 1, half have index 0. Theorem 7.6 implies that the same conclusions hold for $\ell(p,\tau,\cdot;\bar{\xi})$, provided $\tau > 0$ is sufficiently small. Theorem 5.18 and (5.32) then give the first three conclusions. The last conclusion follows as in Theorem 7.6. ∎

Theorems 7.10 and 7.11 resolve Problem 5.20 for the case of a one-parameter perturbation of class β or γ when $\Delta \neq 0$ on appropriate components of G_1. In contrast to the conclusions for a class α perturbation, whether the trivial orbit $G\gamma_0$ breaks is not independent of the material comprising the rod. Rather, (7.11) and Lemma 7.8 indicate that the conclusions generally depend upon the linearly elastic response of the material about the configuration γ_0. If two different materials produce the same linear response about γ_0, and if for that response $\Delta \neq 0$ on appropriate components of G_1, the trivial orbit will break in the same manner for rods composed of the two materials.

When $\Delta = 0$ on the components of G_1 on which $A_{\bar{\xi}}$ is constant, Theorems 7.10 and 7.11 are not applicable. As (6.7) indicates, $h(p,\tau,\cdot;\bar{\xi})$ has degenerate singularities on those components. An analysis of the higher-order terms is then needed to determine if the orbit $G\gamma_0$ breaks under the perturbation. In such a situation, bifurcation will depend upon the nonlinear response of the material comprising the rod.

We close the subsection by illustrating the conclusions of Theorems 7.10 and 7.11 and the remarks concerning the influence of the material on the bifurcation. We construct the examples from the loads which were classified at the end of Section VI.3.

By Lemma 7.8, the components of $\Pi_{2e}\beta(\sigma^{-1}\bar{\xi},\gamma_0)$ and $u_1(p,\sigma;\bar{\xi})$ are polynomials in (x,y) of degree ≤ 1. The following lemmas relate these polynomials and express the coefficients of (7.9) in terms of them.

7.12 Lemma. *On a component of G_1 represent σ by (x,y). Write*

$$u_1(p,\sigma;\bar{\xi}) = \{xa_j(S) - yb_j(S) + c_j(S)\}\hat{\underline{e}}_j \,, \tag{7.12}$$

where we have suppressed the parametric dependence of the coefficient functions on the right hand side upon p and $\bar{\xi}$. Then *the coefficients in (7.9) are given by*

$$\left.\begin{aligned}
\alpha_1 &= \int_I \{A(S)(a'_\alpha(S))^2 - p(a_\alpha(S))^2 + B(S)(a'_3(S))^2\}dS \\
\alpha_2 &= -\int_I \{A(S)a'_\alpha(S)b'_\alpha(S) - pa_\alpha(S)b_\alpha(S) + B(S)a'_3(S)b'_3(S)\}dS \\
\alpha_3 &= \int_I \{A(S)(b'_\alpha(S))^2 - p(b_\alpha(S))^2 + B(S)(b'_3(S))^2\}dS \\
\alpha_4 &= 2\int_I \{A(S)a'_\alpha(S)c'_\alpha(S) - pa_\alpha(S)c_\alpha(S) + B(S)a'_3(S)c'_3(S)\}dS \\
\alpha_5 &= -2\int_I \{A(S)b'_\alpha(S)c'_\alpha(S) - pb_\alpha(S)c_\alpha(S)\ B(S)b'_3(S)c'_3(S)\}dS \\
\alpha_6 &= \int_I \{A(S)(c'_\alpha(S))^2 - p(c_\alpha(S))^2 + B(S)(c'_3(S))^2\}dS,
\end{aligned}\right\} \quad (7.13)$$

for $$A(S) = 2\,\frac{\partial H}{\partial \tau_1}(S,0) \text{ and } B(S) = 2\,\frac{\partial H}{\partial \tau_2}(S,0),$$

as specified in (5.7).

Proof. The lemma follows from (5.6), (5.7), (7.9) and (7.12). ∎

7.13 Lemma. *On a component of* G_1 *represent* σ *by* (x,y). *Write* $\Pi_{2e}\beta(\sigma^{-1}\bar{\xi},\gamma_0) = (\hat{q},\hat{p})$, *for*

$$\left.\begin{aligned}
\hat{q}(S) &= \{xa_{1j}(S) - yb_{1j}(S) + c_{1j}(S)\}\hat{\underline{e}}_j, \\
\hat{p}(S) &= \{xa_{2j}(S) - yb_{2j}(S) + c_{2j}(S)\}\hat{\underline{e}}_j,\quad S = 0,\ 1.
\end{aligned}\right\} \quad (7.14)$$

If $u_1(p,\sigma;\bar{\xi})$ *is given by (7.12), then for* $j = 1, 2, 3$ *fixed,*

a) $a_j(S) \equiv 0$ (*resp.* $b_j(S)$, $c_j(S)$) *if and only if* $a_{1j}(S) \equiv 0$ *and* $a_{2j}(S) \equiv 0$ (*resp.* $b_{1j}(S)$ *and* $b_{2j}(S)$, $c_{1j}(S)$ *and* $c_{2j}(S)$).

b) $c_3(S) \equiv 0$.

c) *Any two coefficients of the components for* $u_1(p,\sigma;\bar{\xi})$ *are linearly dependent if and only if the corresponding coefficients of the components for* $\hat{q}(S)$ *and* $\hat{p}(S)$ *are also.*

Proof.

a) By (5.6) and (5.7), (7.8) specifies a system of three uncoupled (scalar) boundary value problems in the components of $u_1(p,\sigma;\bar{\xi})$ and $\Pi_{2e}\beta(\sigma^{-1}\bar{\xi},\gamma_0)$ relative to $\hat{\underline{e}}_j$. By judiciously choosing (x,y), each problem produces three uncoupled boundary value problems involving the corresponding coefficient functions of the polynomials representing the components of $u_1(p,\sigma;\bar{\xi})$ and $\Pi_{2e}\beta(\sigma^{-1}\bar{\xi},\gamma_0)$. In their weak formulation, these last problems are of one of two forms. Let

$$\mathbb{X} = \{\ f \in W^{r,2}(I,\mathbb{R})\ |\ \int_I f(S)dS = 0\ \}$$

$$\mathbb{Y} = \{\ (g,\underline{h}) \in W^{r-2,2}(I,\mathbb{R}) \times \mathbb{R}^2\ |\ \int_I g(S)dS + h_1 + h_2 = 0\ \}\ .$$

For $(g,\underline{h}) \in \mathbb{Y}$, for $p < p_0$, find $f \in \mathbb{X}$ satisfying either

$$\left.\begin{aligned} &\int_I \Big\{2H_{\tau_1}(S,\underline{0})f'(S)v'(S)-pf(S)v(S)\Big\}dS = \\ &\int_I g(S)v(S)dS+h_1v(1)+h_2v(0) \end{aligned}\right\} \tag{7.15}$$

or

$$\left.\begin{aligned} &\int_I \Big\{2H_{\tau_1}(S,\underline{0})f'(S)v'(S)\Big\}dS = \\ &\int_I g(S)v(S)dS+h_1v(1)+h_2v(0) \end{aligned}\right\}, \tag{7.16}$$

for all $v \in \mathbb{X}$. Here, H_{τ_1} and H_{τ_2} denote the partially differentiated functions occurring in (5.7). Since $p < p_0$, Hypothesis 5.7 and an argument based upon the Fredholm alternative from the theory of elliptic equations identical to the one used in the proof of Proposition 5.8 establishes the existence of a unique solution to either problem (7.15) or (7.16) for each element of $\mathbb{Y}$. Taking $f = a_j$ or $(g,\underline{h}) = (a_{1j},a_{2j})$ give those conclusions of part a) which involve the x-coefficient functions in the polynomials representing the $\hat{\underline{e}}_j$ component of $u_1(p,\sigma;\bar{\xi})$ and $\Pi_{2e}\beta(\sigma^{-1}\bar{\xi},\gamma_0)$. The conclusions involving the other coefficient functions follow similarly.

b) A computation shows that the $\underline{e}_1 \otimes \underline{e}_2$ and $\underline{e}_2 \otimes \underline{e}_1$ components of either tensor function comprising $\sigma^{-1}\bar{\xi}$ is a homogeneous polynomial of degree 1 in (x,y). So the $\hat{\underline{e}}_3$ component of the tensor functions comprising $\Pi_{2e}\beta(\sigma^{-1}\bar{\xi},\gamma_0)$ are homogeneous polynomials of degree 1 in (x,y). Hence $(c_{13},c_{23}) \equiv 0$, and the conclusion follows from part a).

c) This part follows from the uniqueness of the solution to (7.15) or (7.16). ∎

We now present the examples.

7.14 Example. *An Inhomogeneous Non-Parallel Force Distribution.* Consider the load of Example 6.23: $\bar{\xi}_2 = (\mu_2,0)$,

$$\mu_2(S) = \pi R^3[\underline{e}_1 \otimes \underline{e}_1 - \frac{\pi}{2}\sin(\pi S)\ \underline{e}_2 \otimes \underline{e}_2]. \tag{7.17}$$

Then $\bar{\xi}_2$ is a canonical form of class β_2. Let $\sigma \in SG_1$ be represented by (x,y). Then (3.35), (3.39), and Lemma 5.3 imply that $\Pi_{2e}\beta(\sigma^{-1}\bar{\xi}_2,\gamma_0) = (\hat{q}_2,0)$, for

$$\hat{q}_2(S) = \{(-y)(-\pi R^3)(1 - \frac{\pi}{2}\sin(\pi S)\ )\}\hat{\underline{e}}_3.$$

By Lemma 7.13 and (7.12), for $b_3(S)$ depending parametrically on p and $\bar{\xi}_2$,

$$u_1(p,\sigma;\bar{\xi}_2) = -yb_3(S)\hat{\underline{e}}_3.$$

By (5.11),(7.11) through (7.13), and $p < p_0$,

$$\alpha_1 = \alpha_2 = \alpha_4 = \alpha_5 = \alpha_6 = 0, \text{ and } \alpha_3 > 0.$$

Hence, $\Delta = \alpha_3^{\,6} > 0$, and Theorem 7.10 implies there will be six equilibrating configurations, three of which are stable. Four of the equilibrating configurations may be characterized as lying on SG_1 near (1,1), $(R_{\pi/2},1)$, $(R_\pi,1)$, and $(R_{3\pi/2},1)$. The other two lie on the other component of G_1 near (J,1) and $(R_\pi J,1)$ (See Figure 7.3.1).

Remarks.

1. For this example the results on SG_1 can be confirmed by directly computing the critical points for $Q(x,y) = \alpha_3 y^2$.

2. In contrast to the case of a class α perturbing load, we cannot simply examine the net torque produced by $\bar{\xi}_2$ to deduce which configurations are equilibrating and which are not. Computing $k(\bar{\xi}_2, R_\phi \gamma_0)$ shows that $\bar{\xi}_2$ produces no net torque in any of the configurations $R_\phi \gamma_0$. It signifies that SG_1 is contained in the critical manifold for $\bar{\xi}_2$, the object which arose in the first order analysis. It does not discern the critical points on SG_1 arising from the higher order terms in the expansion of ℓ.

3. The symmetry of $\bar{\xi}_2$ is noteworthy. By (7.17), the subgroup of Γ generated by (R_π, R_π) and $(J\Sigma_3, J\Sigma_3)$ leaves $\bar{\xi}_2$ invariant. By Lemma 4.28 and (5.31), if $\sigma = (R_\phi, 1)$ is a critical point, so are $(R_{-\phi}, 1)$, $(R_{\phi+\pi}, 1)$, and $(R_{-\phi+\pi}, 1)$. However, symmetry alone is insufficient to distinguish the existence of four solutions on SG_1, instead of two. Take $\phi = 0$, for example. Hence, Δ gives information other than that contained in the symmetry of $\bar{\xi}_2$.

4. For this particular load $\alpha_4 = \alpha_5 = 0$. So, $\Delta > 0$, regardless of the material comprising the rod. Hence, the bifurcation conclusions are independent of the material comprising the rod.

7.15 Example. *A Constant Gravitational Force Distribution.* Consider the load of Example 6.23: $\bar{\xi}_4 = (\mu_4, 0)$,

$$\mu_4(S) = -\pi R^2 (1 - S) \underline{e}_2 \otimes \underline{e}_3. \tag{7.18}$$

It is a canonical form of class γ. For $\sigma \in G_1$, a computation gives $\Pi_{2e}\beta(\sigma^{-1}\bar{\xi}_4, \gamma_0) = (\hat{q}_4, 0)$,

$$\hat{q}_4(S) = (\pi R^2)(1/2 - S)(y\hat{\underline{e}}_2 + x\hat{\underline{e}}_1).$$

By Lemma 7.13,

$$u_1(p, \sigma; \bar{\xi}_4) = x a_1(S)\hat{\underline{e}}_1 - y b_2(S)\hat{\underline{e}}_2,$$

where $a_1(S) = -b_2(S)$. Equations (7.13) and (5.11) imply that

$$\alpha_2 = \alpha_4 = \alpha_5 = \alpha_6 = 0,$$

$$\alpha_1 = \alpha_3 > 0.$$

So $\Delta = 0$ on each component of G_1, and Theorem 7.11 is inapplicable in this case.

However, some observations can be made about this case. The results for the coefficients indicate that (7.9) is constant on G_1. If the material comprising the body were linearly elastic, (6.7) would be exact at order τ^2, and $\ell(p,\tau,\cdot;\bar{\xi}_4)$ would be constant on G_1. Consequently, no bifurcation of the orbit $G\gamma_0$ would occur. If the material comprising the rod had a nonlinear response, no conclusions could be drawn. So the breaking of the orbit depends upon the nonlinear nature of the material comprising the rod.

7.16 Example. *An Inhomogeneous Distribution of Moments*
In the notation of Section III.2, take $l_{3D} = (0,\tau)$, where

$$\tau(X^1,X^2,S) = \begin{cases} \sin(2\pi S)X^2\underline{e}_3 & (X^1,X^2,S) \in \partial\mathcal{B}(S),\ S \neq 0,\ 1 \\ 0 & (0,0,S),\ S = 0,\ 1. \end{cases}$$

Averaging over the cross section gives a characterization of the force distribution as a load in the special Cosserat theory gives $l = (f,\mathbb{Q},t,\mathbb{P}) = (0,\mathbb{Q},0,0)$,

$$\mathbb{Q}(S) = \pi R^3\sin(2\pi S)\underline{e}_3 \otimes \underline{e}_2.$$

The load produces an inhomogeneous distribution of moments parallel to the $\underline{e}_1$ axis along the centerline of the rod. By (3.29) we may characterize the load in the Kirchhoff theory as $\bar{\xi}_5 = (\mu_5,0)$,

$$\mu_5(S) = \pi R^3\sin(2\pi S)\underline{e}_3 \otimes \underline{e}_2. \qquad (7.19)$$

The load is a canonical form of class γ. For $\sigma \in G_1$, a computation gives $\Pi_{2e}\beta(\sigma^{-1}\bar{\xi}_5,\gamma_0) = (\hat{q}_5,0)$,

$$\hat{q}_5(S) = -\pi R^3 \sin(2\pi S)\hat{\underline{e}}_1.$$

By Lemma 7.13,

$$u_1(p,\sigma;\bar{\xi}_5) = c_1(S)\hat{\underline{e}}_1.$$

By (7.11) through (7.13), $\Delta = 0$ on G_1, so Theorem 7.11 is inapplicable in this case. However, (7.13), (7.9), (6.7) and (6.5) show that $\ell(p,\tau,\cdot;\bar{\xi}_5)$ is constant to third order in τ on G_1, so comments similar to those made at the end of the last example also hold for this example.

In contrast to the previous examples the superposition of the previously given loads produce a perturbation for which the bifurcation of the orbit $G\gamma_0$ can change with the linear elastic response of the material comprising the rod.

7.17 Example. *A Combination of Perturbing Loads*
Let $\bar{\xi} = \bar{\xi}_2 + \bar{\xi}_4 + \bar{\xi}_5$, where the loads on the right hand side are given by (7.17) through (7.19). A computation shows that $\bar{\xi}$ is a canonical form of class β_2. Moreover, for $\sigma \in SG_1$, $\Pi_{2e}\beta(\sigma^{-1}\bar{\xi},\gamma_0) = (\hat{q},0)$, where $q = \hat{q}_2 + \hat{q}_4 + \hat{q}_5$. By Lemma 7.13,

$$u_1(p,\sigma;\bar{\xi}) = (xa_1(S) + c_1(S))\hat{\underline{e}}_1 - yb_2(S)\hat{\underline{e}}_2 - yb_3(S)\hat{\underline{e}}_3,$$

and $a_1(S) = -b_2(S)$. By (7.13) and (5.11),

$$\alpha_1 - \alpha_3 = -\int_I B(S)(b_3'(S))^2 dS < 0,$$

$$\alpha_4 = \int_I \{A(S)a_1'(S)c_1'(S) - pa_1(S)c_1(S)\}dS \neq 0,$$

$$\alpha_2 = \alpha_5 = 0.$$

So on SG_1, $\Delta > 0$, $\Delta = 0$, or $\Delta < 0$ is possible, depending on how $A(S)$ and $B(S)$ are related. Hence under this perturbation, Theorem 7.10 implies that if

$$\left| \frac{\alpha_4}{2(\alpha_1 - \alpha_3)} \right| \neq 1,$$

the rod will exhibit either four or six equilibrating configurations nearby $G\gamma_0$, with either two or four of them continuously deformable from the trivial configuration γ_0. The number depends upon the linearly elastic response of the material comprising the rod. Half of them will be stable equilibria.

Remarks.

1. For this case the conclusions on SG_1 may be confirmed by directly computing the critical points for $Q(x,y) = \alpha_1 x^2 + \alpha_3 y^2 + \alpha_4 x$, where $\alpha_3 > \alpha_1 > 0$, and $\alpha_4 \in \mathbb{R}$.

2. By (7.13) and the previous examples, $\Delta = 4(\mathcal{B}^2 - \mathcal{C}^2)^3$, where

$$\mathcal{B} = \int_I B(S) b_3' b_3' dS > 0,$$

$$\mathcal{C} = \int_I [A(S) a_1 c_1 - p a_1 c_1] dS \in \mathbb{R},$$

and A and B are as specified in Lemma 7.12. Since A is a measure of the bending stiffness of the rod about the trivial configuration, and B is a measure of its twisting stiffness, the results indicate that a rod composed of a material which makes it easier to bend than to twist may exhibit four equilibrating configurations when subjected to the perturbing load $\bar{\xi}$, whereas one easier to twist than to bend may exhibit only two.

3. For this example, all three perturbing loads are needed to obtain the variety of behavior indicated for different materials comprising the body. If $\hat{q}_4$ or $\hat{q}_5$ were absent, $\alpha_4 = 0$. Hence, four equilibrating configurations would result, regardless of the material comprising the rod. If $\hat{q}_2$ were absent, $\alpha_1 = \alpha_3$; hence, two equilibrating configurations would result, regardless of the material.

Figures 7.3.1 and 7.3.2 illustrate the conclusions of the branching theorem and the subsequent remarks for Example 7.14. Figure 7.3.1 depicts the three-dimensional force distribution acting on a single cross section of the rod in various rotations $\sigma\gamma_0$ of the trivial configurations. As indicated in the remarks the equilibrating ones are not obvious from the diagrams. However, since $\Delta > 0$, there are four of them, illustrated in Figure 7.3.2 by the branching diagram on $SG_1 \times \mathbb{R}$.

Illustration of the Branching Theorem

Case β

Example 7.19 $\xi_1 = (\pi R^3(\underline{e}_1 \otimes \underline{e}_1 - (\pi/2)\sin(\pi S)\underline{e}_2 \otimes \underline{e}_2), 0)$

S fixed:

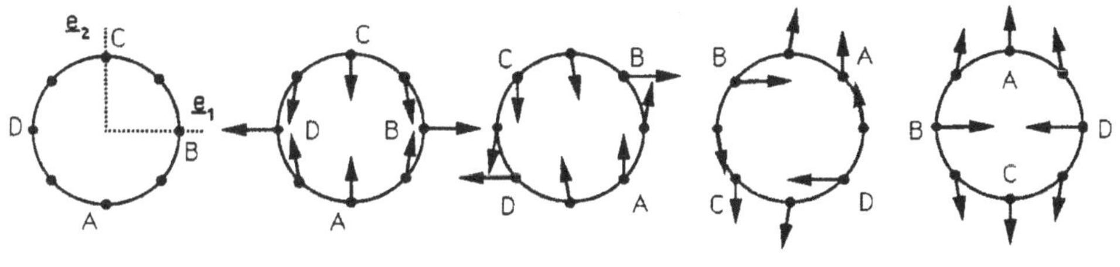

eference	γ_0	$\sigma_1\gamma_0$	$\sigma_2\gamma_0$	$\sigma_3\gamma_0$
quilibrium	yes	?	?	yes

Figure 7.3.1

Illustration of the Branching Theorem (continued)

$\Delta > 0$ *Branching Diagram in* $SG_1 \times \mathbb{R}$

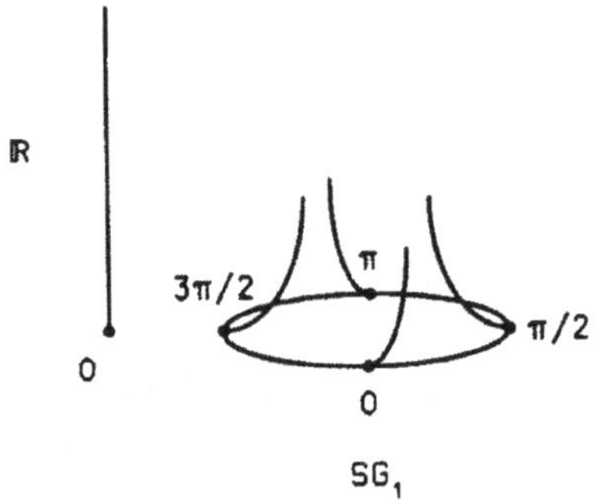

Figure 7.3.1

VIII. CONCLUSIONS AND ADDITIONAL PROBLEMS

In summary we've illustrated that we may use the techniques of the theory of singularities to examine how an orbit of equilibrating configurations for a rod subject to a symmetric load alters when you perturb the load in a way that breaks the symmetry. We did so by examining in detail what happens to the orbit generated by the straight configuration for a rod in the Kirchhoff theory which has a circular cross section, is composed of an isotropic material, and is initially subjected to an axially symmetric compressive load. Under suitable assumptions we showed that the problem may be formulated as a problem of finding the singularities of a potential function (Problem 4.25). Next, we showed that this problem may be reduced to one specified on a finite dimensional space in a manner which depends upon the pressure of the compressive load (Theorems 5.18 and 5.24). We completed the reduction for that case where the pressure approaches the first value at which buckling occurs (Problem 5.19). We then resolved the bifurcation problem for this case by classifying the admissible perturbing loads (dead loads) into three categories (Theorem 6.20), and by determining how the orbit of equilibrating configurations for the unperturbed problem alters upon the application of each type of perturbing load (Theorems 7.6, 7.10 and 7.11). One type altered the orbit in a manner which was independent of the material comprising the rod. The other types could alter the orbit, but the alteration depended upon the response of the material comprising the rod. Thus, we could see mathematically how the alteration of the orbit depended upon the extent to which the perturbing load broke the original symmetry, and upon the material comprising the rod.

This work represents but a first step in the analysis of the bifurcation problem of interest. Several questions immediately present themselves.

The results of Section 7 (Theorems 7.6, 7.10, and 7.11) indicate when knowledge of the response of the material comprising rod up through order two determines how the orbit of equilibrating configurations to the symmetrically loaded problem alters under perturbation. However, Examples 7.15 and 7.16 indicate that this knowledge does not completely resolve the problem when $p < p_0$. Rather, there are instances where knowledge of the response of the material beyond order two is needed to determine how the orbit alters. Can a theory of singularities of higher order be used to refine the classification of Theorem 6.20 and augment the conclusions of Theorems 7.10 and 7.11? Here the isotropy group of $\bar{\xi}$, the subgroup of Γ which leaves $\bar{\xi}$ invariant, becomes important.

A second question is what happens when the pressure reaches and goes beyond the first buckling value? As the work of [9] shows, equilibrating configurations other than the straight configuration arise for the symmetrically loaded problem. Each one of them generates an orbit of equilibrating configurations. How do these orbits alter when you perturb the load in an asymmetric way? For this question Theorem 5.24 becomes central to the analysis of the bifurcation problem. The theorem produces a reduced bifurcation problem over the three-dimensional space $G_1 \times U_\varepsilon$ (see (5.36)). Little can be said about this problem in general. However, in Problem 5.26 with $\bar{\xi}$ of class α, for $\tau > 0$ a second reduction is possible, leading to a bifurcation problem over the two-dimensional space U_ε. Even in this case, the problem cannot be entirely resolved. The isotropy group of $\bar{\xi}$ becomes significant. For example, we conjecture that if the isotropy group of $\bar{\xi}$ is the dihedral group D_4, then we can "recognize" the reduced problem as an unfolding of a double cusp rendered nondegenerate by the breaking of the O(2) symmetry by $\bar{\xi}$. For such an unfolding, secondary bifurcations emerge at $p = p_0$ (see [23], §10).

We can contrast this conjecture at $p = p_0$ with the results we would obtain for the bifurcation problem for the symmetrically loaded rod with ellipsoidal cross section. Adapting the work of E. Buzano in [35], we are again lead to unfold a double cusp, now rendered nondegenerate by the breaking of the isotropy of the material response by the ellipsoidal geometry of the cross section. By contrast, we maintain the isotropy of the material response in the conjecture; the nondegeneracy results from the breaking of the symmetry by the perturbing load.

It is instructive to compare Problem 5.25 with the analogous problem for the prismatic rod in the Kirchhoff theory. Here the work of [4] [23], and [35] indicate that the effects of a symmetry breaking perturbing load like $\bar{\xi}$ can be fully described by a universal unfolding of a nondegenerate double cusp. There is a reason we can resolve the problem for the prismatic rod, but must step gingerly around Problem 5.25. It lies in the fact that the perturbing load $\bar{\xi}$ is breaking a *finite* group of symmetries for the prismatic rod, and a *continuous* group of symmetries (O(2)) for the cylindrical rod. Ideally, we would like to view (5.36) as an unfolding of f_2 evaluated at $\lambda = 0$, $\bar{\xi} = 0 \in \mathcal{K}$. However, the analysis leads to the unfolding of a *degenerate* double cusp having infinite codimension, even when perturbing loads are required to maintain finite groups of symmetry (see [23], §10). Some promise towards resolving this difficulty lies in the ideas of M. Golubitsky and D. Schaeffer in [36] and in the work of J. Damon in [37] and [38].

The conclusions for the bifurcation problem presented in this work depend heavily upon choosing the perturbing loads to be *dead* loads in the sense of Assumption 3.2. We may consider perturbations by load systems other that (3.4). For example, if we replace (3.4) by

$$\left.\begin{aligned} f_{\chi}(S) &\equiv \gamma(S)f(S) \\ \mathbb{Q}_{\chi}(S) &\equiv \gamma(S)\mathbb{Q}(S), \quad 0 \le S \le 1, \\ t_{\chi}(S) &\equiv \gamma(S)t(S) \\ \mathbb{P}_{\chi}(S) &\equiv \gamma(S)\mathbb{P}(S), \quad S = 0, 1, \end{aligned}\right\} \tag{8.1}$$

we will be describing a load system on a rod in the special Cosserat theory which may be viewed as arising from a three-dimensional force distribution which changes with the three-dimensional configuration of the rod (a *follower* load). In particular, if the perturbing loads on the rod arise from a distribution of pressure on its surface in three dimensions, then the corresponding load in the special Cosserat theory would be of the form (8.1). How are the orbits of the equilibrating configurations for the rod in the Kirchhoff theory altered when the rod is perturbed by loads of the form of (8.1)? It may be that an analysis of the reduced potential function (5.30) to order one or two is inadequate to resolve the bifurcation problem for this load system. An analysis using a theory of higher order singularities may be necessary.

Throughout this work we assumed that the material comprising the rod was hyperelastic (Hypothesis 4.3). Is it possible to analyze the problem of interest if the material comprising the rod were Cauchy elastic, as opposed to hyperelastic? As indicated in Section V.2, Problem 5.19 and Theorem 5.9 do not depend upon our variational formulation of the equilibrium problem (see [2], p. 371). So we can formulate the bifurcation problem and reduce it to one in a finite dimensional setting. However, in the absence of a reduced potential function like (5.30) the techniques of the theory of singularities for real-valued functions which were used in this work to produce Theorems 7.6, 7.10, and 7.11 would no longer be applicable. Rather, the techniques of the full theory of singularities of mappings and the bifurcation theory would have to be employed.

Other problems involving rods in the Kirchhoff theory suggest themselves. Can these techniques be used to analyze the problem analogous to the one presented here for the rod whose natural configuration is a helix (a coiled spring)? Is it possible to analyze the problem where there is no natural configuration for the rod, as when the rod possesses a residual stress?

Finally, it would be of value to extend the work done here and suggested above to rods in the special Cosserat theory and the Cosserat theory in general (see [33]). The reader is invited to consider such an undertaking.

REFERENCES

[1] Chillingworth, D. R. J., Marsden, J. E., and Wan, Y. H., Symmetry and Bifurcation in Three-Dimensional Elasticity, Part I. *Archive for Rational Mechanics and Analysis*, **80**, 1982, 295-331.

[2] Chillingworth, D. R. J., Marsden, J. E., and Wan, Y. H., Symmetry and Bifurcation in Three-Dimensional Elasticity, Part II. *Archive for Rational Mechanics and Analysis*, **83**, 1983, 362-395.

[3] Wan, Y. H., and Marsden, J. E., Symmetry and Bifurcation in Three-Dimensional Elasticity, Part III. *Archive for Rational Mechanics and Analysis*, **84**, 1983, 203-233.

[4] Buzano, E., Geymonat, G., and Poston, T., Post-Buckling Behavior of a Non-Linearly Hyperelastic Thin Rod with Cross-Section Invariant under the Dihedral Group D_n. *Archive for Rational Mechanics and Analysis*, **89**, 1985, 307-388.

[5] Antman, S., and Jordan, K., Qualitative Aspects of the Spatial Deformation of Nonlinearly Elastic Rods. *Proceedings of the Royal Society of Edinburgh*, **73A**, 1975, 85-105.

[6] Antman, S., Ordinary Differential Equations of One-Dimensional Nonlinear Elasticity I: Foundations of the Theories of Nonlinearly Elastic Rods and Shells. *Archive for Rational Mechanics and Analysis*, **61**, 1976, 307-351.

[7] Abraham, R., and Marsden, J. E., *Foundations of Mechanics*. Second Edition. Reading, The Benjamin-Cummings Publishing Company, 1978.

[8] Palais, R., *The Foundations of Global Nonlinear Analysis*. Reading, The Benjamin-Cummings Publishing Company, 1968.

[9] Antman, S.., and Kenney, C., Large Buckled States of Nonlinearly Elastic Rods under Torsion, Thrust, and Gravity. *Archive for Rational Mechanics and Analysis,* **76**, 1981, 289-338.

[10] Maddocks, J. H., Stability of Nonlinearly Elastic Rods. *Archive for Rational Mechanics and Analysis,* **85**, 1984, 311-354.

[11] Ericksen, J. L., and Truesdell, C., Exact Theory of Stress and Strain in Rods and Shells. *Archive for Rational Mechanics and Analysis,* 1, 1958, 295-323.

[12] Cohen, H., A Nonlinear Theory of Elastic Directed Curves. *International Journal of Engineering Sciences,* **4**, 1966, 511-524.

[13] Wang, C-C., and Truesdell, C., *Introduction to Rational Elasticity.* Leyton, Noordhoff International Publishing Company, 1973.

[14] Smale, S., An Infinite Dimensional Version of Sard's Theorem. *American Journal of Mathematics,* **87**, 1965, 861-866.

[15] Pierce, J. F., Global Models for Cosserat Continua and some Fibrations of Palais, Cerf, and Smale. *Physical Mathematics and Nonlinear Partial Differential Equations,* J. Lightbourne, III, and S. Rankin, III, editors. New York, Marcel Dekker, Inc., 1985, 239-257.

[16] Golubitsky, M., and Guillemin, V., *Stable Mappings and their Singularities.* Graduate Texts in Mathematics # 14. New York, Springer-Verlag, Inc., 1973.

[17] Epstein, M., Elzanowski, M., and Śniatycki, J., Locality and Uniformity in Global Elasticity. *Differential Geometric Methods in Mathematical Physics,* H. Doebner, and J. Hennig,

editors. Lecture Notes in Mathematics # 1139. New York, Springer-Verlag Inc., 1985, 300-310.

[18] Husemuller, D., *Fiber Bundles.* McGraw-Hill Series in Higher Mathematics. New York, McGraw-Hill Book Company, 1966.

[19] Antman, S., Monotonicity and Invertibility Conditions in One Dimensional Nonlinear Elasticity. *Symposium on Nonlinear Elasticity,* R. W. Dickey, editor. New York, Academic Press, Inc., 1973, 57-92.

[20] Eells, J., Elliptic Operators on Manifolds. *Global Analysis and its Applications, Volume I.* Proceedings of the International Seminar Course, Trieste 1972. International Center for Theoretical Physics, Trieste, Vienna. New York, International Atomic Energy Agency, 1974.

[21] Treves, F., *Basic Linear Partial Differential Equations.* New York, Academic Press, Inc., 1975.

[22] Hartman, P., *Ordinary Differential Equations.* New York, John Wiley and Sons, Inc., 1964.

[23] Golubitsky, M., and Schaeffer, D., *Singularities and Groups in Bifurcation Theory, Volume I.* New York, Springer-Verlag, Inc., 1985.

[24] Sattinger, D. H., *Branching in the Presence of Symmetry.* CBMS-NSF Regional Conference Series in Applied Mathematics, Volume 40. Philadelphia, Society for Industrial and Applied Mathematics, 1983.

[25] Lu, Y-C., *Singularity Theory and an Introduction to Catastrophe Theory.* New York, Springer-Verlag, Inc., 1976.

[26] Milnor, J., *Morse Theory.* Princeton, Princeton University Press, 1963.

[27] Mather, J., Stability of C^{∞} Mappings II: Infinitesimal Stability implies Stability. *Annals of Mathematics*, **89**, 1969, 254-291.

[28] Elliason, *Journal of Differential Geometry*, **1**, 1967, 169-194.

[29] Helgason, S., *Differential Geometry and Symmetric Spaces.* New York, Academic Press, 1962.

[30] Golubitsky, M., and Schaeffer, D., Imperfect Bifurcation in the Presence of Symmetry. *Communications in Mathematical Physics*, **67**, 1979, 205-232.

[31] Golubitsky, M., and Stewart, I., Hopf Bifurcation in the Presence of Symmetry. *Archive for Rational Mechanics and Analysis*, **87**, 1985, 107-165.

[32] Abraham, R., and Robbin, J., *Transversal Mappings and Flows.* New York, W. A. Benjamin, Inc., 1967.

[33] Simo, J., Marsden, J., and Krishnaprasad, P., The Hamiltonian Structure of Nonlinear Elasticity. *Archive for Rational Mechanics and Analysis*, **104**, 1988, 125-184.

[34] Vanderbauwhede, A., *Local Bifurcation and Symmetry.* Research Notes in Mathematics No.73. Boston, Pitman Publishing, Inc., 1982.

[35] Buzano, E., Secondary Bifurcations of a Thin Rod under Axial Compression. *SIAM Journal of Mathematical Analysis*, **17**, 1986, 312-321.

[36] Golubitsky M., and Schaeffer, D., A discussion of Symmetry and Symmetry Breaking. *Singularities: Proceedings of the Symposia in Pure Mathematics.* **40.1**, 1983, 499-516.

[37] Damon, J., The Unfolding and Determinacy Theorems for Subgroups of $\mathcal{A}$ and $\mathcal{K}$. *Singularities: Proceedings of the Symposia in Pure Mathematics.* **40.1**, 1983, 233-254.

[38] Damon, J., The Unfolding and Determinacy Theorems for Subgroups of $\mathcal{A}$ and $\mathcal{K}$. *Memoirs of the American Mathematical Society.* **306**, 1984.

INDEX

M

O

P

R

S